Modern Concepts of Landscape Designing

THE AUTHORS

Dr. Suneeta Singh is presently serving as Assistant Professor and Head at the Department of Horticulture, School of Agricultural Sciences, Shri Guru Ram Rai University, Dehradun, Uttarakhand. She is associated with teaching and research in the courses pertaining to horticulture for the last 8 years. In the continuing service tenure, she has worked on landscape gardening, orchard management aspects and protected cultivation of many horticultural crops. She has acquired her Graduation in Agriculture from Banaras Hindu University, Varanasi, U.P.; Post graduation and Ph.D. in Horticulture from G. B. Pant University of Agriculture & Technology, Pant nagar, Uttarakhand. The author was recipient of Mandi Parishad merit scholarship in B.Sc. (Ag.) and ICAR-JRF Fellowship in M.Sc. She also qualified ICAR-NET. Dr Singh started her professional career in 2010 after earning her Ph.D. degree as SRF in Department of Vegetable Science, G.B.P.U.A. &T., Pant nagar. Dr Singh was also worked as Research Associate in Vivekananda Parvatiya Krishi Anusandhan Sansthan (VPKAS), ICAR, Almora, Uttarakhand. She has published 1 Book, 6 Book chapters, 14 Popular articles covering various aspects of Horticultural technologies and more than 38 Research papers in peer reviewed NAAS rated scientific journals of International and National reputes. In recognition of her services to the scientific journal, the author was selected as Review Board Member by Acta Scientific Agriculture. Dr Singh successfully guided 6 students leading to PG degree in Horticulture.

Dr. Anil Kumar Saxena is currently working as Associate Professor, Department of Soil Science, School of Agricultural Sciences, Shri Guru Ram Rai University, Dehradun, Uttarakhand. Dr. Saxena has more than 12 years of working experience in teaching, research and extension. He has Post graduated and also acquired Doctorate in Soil Science from G.B. Pant University of Agriculture & Technology, Pantnagar, Uttarakhand. He has qualified ICAR-NET in same subject. Dr Saxena has 8 years of administrative experience as Principal/Dean Incharge in School of Agricultural Sciences, SGRRU, Dehradun. Dr. Saxena successfully guided 16 students leading to PG degree. Dr. Saxena has to his credit 32 research papers, 1 research bulletin, 1 book, 5 book chapters, 12 popular articles, radio talks and technical folders. He is also involve as resource person in organic farming and organic certification process under SGRR University, Dehradun. He has handled and completed one UGC funded project. Dr Saxena has organised various trainings in field of Agriculture and also worked as convener of Second State Science and Technology seminar organised by Uttarakhand Science Education and Research Centre (USERC), Dehradun in 2019.

Modern Concepts of Landscape Designing

Suneeta Singh
Anil Kumar Saxena

2026
Daya Publishing House®
A Division of
Astral International Pvt. Ltd.
New Delhi – 110 002

Reprinted 2026

Publisher's Note:

Every possible effort has been made to ensure that the information contained in this book is accurate at the time of going to press, and the publisher and author cannot accept responsibility for any errors or omissions, however caused. No responsibility for loss or damage occasioned to any person acting, or refraining from action, as a result of the material in this publication can be accepted by the editor, the publisher or the author. The Publisher is not associated with any product or vendor mentioned in the book. The contents of this work are intended to further general scientific research, understanding and discussion only. Readers should consult with a specialist where appropriate.

Every effort has been made to trace the owners of copyright material used in this book, if any. The author and the publisher will be grateful for any omission brought to their notice for acknowledgement in the future editions of the book.

Cataloging in Publication Data--DK
Courtesy: D.K. Agencies (P) Ltd. <docinfo@dkagencies.com>

Singh, Suneeta *(Assistant professor of horticulture),* **author.**
Modern concepts of landscape designing / Suneeta Singh and Anil Kumar Saxena.
pages cm
Includes bibliographical references.
ISBN : 978-93-90435-26-5 (HB)

1. Landscape design. 2. Gardens--Design. 3. Landscape gardening. I. Saxena, Anil Kumar (Associate professor of soil science), author. II. Title.

LCC SB473.S56 2021 | DDC 635 23

Published by : **Daya Publishing House®**
A Division of
Astral International Pvt. Ltd.
– ISO 9001:2015 Certified Company –
4736/23, Ansari Road, Darya Ganj
New Delhi-110 002
Ph. 011-43549197, 23278134
E-mail: info@astralint.com
Website: www.astralint.com

Digitally Printed at : **Replika Press Pvt. Ltd.**

Preface

Landscape designers work on a canvas that is distinctly different from other art forms. The "art" is always changing as the plants grow, environmental conditions change, and people use the space. For this reason, landscape designers use a design process that systematically considers all aspects of the land, the environment, the growing plants, and the needs of the user to ensure a visually pleasing, functional, and ecologically healthy design. A landscape is made good or bad by the way in which these components are both selected and are arranged together. The landscape is constantly changing, and a good designer must foresee and account for changes which are likely to occur.

Encompassing 14 chapters, the book is a detailed and comprehensive summary of all the possible facts and figures regarding modern concepts of landscape designing. However, the things are not only limited to the ecosphere that are known to mankind. There may be the chances of lacks and lacuna. Thus scope for further improvement always rests. I shall be highly obliged to all readers who will be extending their fellow-feeling in future improvement of the present endeavour.

We feel indebted to my teachers from Department of Horticulture, Institute of Agricultural Sciences, BHU, Varanasi as well as from Department of Horticulture, G.B. Pant University of Agriculture and Technology, Pant nagar who ushered we the tottering talk of the subject and it is due to their blessings that we have been able to get the manuscript compiled. We don't find words sufficient to express our indebtedness to our loveable parents, brother, sisters, brother-in-laws and sister-in-laws, whose blessings and inspirations enforced us to complete this project. We wish to place on record my sincere thanks to my husband Dr. Anil, my wife Dr Suneeta and our son Mr. Avyaan for bearing us in keeping us free for writing work. We express our gratefulness to all those whose deeds have been useful directly or indirectly in this compilation. We are very much confident that this book would find a place of preference among various categories of readers in Agriculture as well as Horticulture.

SUNEETA SINGH

ANIL KUMAR SAXENA

Contents

Chapter 1

Introduction to Landscape Gardening

The designing and systematic planning of a garden is an art. A garden architect should learn enough of geology, geography, garden history, styles of gardening, and above all one should have a profound knowledge about plants, soils, land topography and local environmental conditions.

Landscape

"A landscape may be defined as any area, either big or small, on which it is possible or desirable to mould a view or a design".

Landscape gardening or Landscape designing or Landscape architecture

Landscape gardening may be described as the application of garden forms, methods, and materials with a view to improving the landscape. The art of designing is known as "landscape architecture," although the older term "Landscape gardening" is also popular.

Recommendations for implementing different principles of gardening

- Do not imitate another garden for designing.
- One has to develop one's own design giving due consideration to the local environmental conditions.
- Avoid overcrowding of many plant species in a limited area.
- Take advantage of natural topography while designing the garden.
- In each garden there should be at least one feature or if there be a second, then these two should harmonize with each other.
- Before planning a design one must be sure for what purpose the garden is made i.e. for utility or beauty or both.

Art Principles of Landscape Gardening

1. Initial Approach

- A good designer should design the landscape in the available space or land area.
- The natural topography should be retained as much as possible.

- Fencing should be such that it looks natural as far as practicable and it should not obstruct any natural view. For example, if there is natural forest scenery or a hillock just outside the boundary, it should be incorporated in the garden design in a thoughtful manner so that it appears to be a part of the landscape designing.

2. Axis

- It is an imaginary line in any garden around which the garden is created striking a balance.
- In a formal garden, the central line is the axis.
- At the end of an axis, generally there will be a centre of attraction, although other architectural features such as bird-bath or sundial can also be erected at about the midpoint.

3. Focal Point

- A focal point in every garden is a centre of attraction which is generally an architectural feature focused as a point of interest such as statue, fountain, rockery, specimen tree etc.

4. Mass Effect

- The use of single plant species in large numbers in one place is done to have mass effect.
- One should see that such mass arrangements do not become monotonous i.e. the sizes of masses should be varied.

5. Unity

- Unity in a garden is very important and it will improve the artistic look of the garden.
- Unity has to be achieved from various angles.
- First, the unity of style, feeling, and function between the building and the garden has to be achieved.
- Secondly, the different components of the gardens should merge harmoniously with each other. The aim is to create an overall impression of the garden rather than blowing up some special features.

- Lastly, it is of prime importance to achieve harmony between the landscape outside and the garden. A garden laid out in complete defiance of the local conditions may look exotic, but is not a successful garden.
- As for example, cacti planted in a seashore garden is completely out of place as these are inhabitants of dry localities.
- To achieve a unity between the building and the garden it is a common practice to train creepers on the front porch which cover the rudeness of the masonry work and also bring the building closer to nature. For the same reasons, foundation plantings are also done.
- A foundation planting broadly means the planting of bushy plants near the foundation of the building.

Unity using different colours

Coarse Texture Plants

Creation of mobility by live features

6. Space

- The aim of every garden design should be such that the garden should appear larger than its actual size.
- One way of achieving this is to keep vast open spaces, preferably under lawn and restrict the plantings in the periphery, normally avoiding any planting in the centre.
- But if any planting has to be done in the centre the choice should be a tree which branches at a higher level on the trunk (or the lower branches are removed), and not a bushy shrub.
- Such planting will not obstruct the view or make the garden appear smaller than its size.
- Another suggestion to create the illusion of more space in a large public garden is to alternate large lawns followed by a group of trees. A large open space planted haphazardly all over with trees looks smaller than its size.
- The techniques of creating an illusion of more space are also referred as 'Forced Perspective.

7. Divisional Lines

- In a landscape garden, there should not be any hard and fast divisional lines. However, there is the necessity of dividing or rather screening a dustbin or vermicompost pit or a Gardner's quarter or a vegetable garden from the rest of the garden.
- In fact areas under lawn, gravel, stone or cement path, and shrubbery border have their natural divisional lines from its immediate neighbour though these are not discreet.
- The divisional lines should be artistic with gentle curves and these should also be useful.
- Above all these lines should harmonize with one another.

8. Proportion and Scale

- Proportion in a garden may be defined as a definite relationship between different elements.
- For example, a rectangle having a ratio of 5:8 is considered to be of pleasing proportion.
- As this ratio comes down the form looks neither a square nor a rectangle and the design becomes undesirable.
- There is no set rule with respect to scale and proportion in a garden, ultimately the design should look pleasant. It is better to have an ad-hoc design first and then try it out on the actual spot.
- If the design looks appealing as well as pleasing, it is implemented. When a shrubbery border has to be planted the outer design is marked by arranging a rubber hose or thick wet rope in different designs on the spot and the one, which looks best, is adopted.
- Judgment of scale and proportion fully depends on the individual experience and the thorough knowledge of plants.

Balance

Colour and Tone

Axis

9. Texture

- The surface character of a garden unit is referred to as texture. The texture of the ground, the leaves of a tree or shrub will all determine the overall effect of the garden.

- Generally, the texture is of three types viz., Fine, Medium and Course. The texture of plant depends on its leaf size, arrangement of branches and compactness of canopy.
- Moreover, the texture can also be classified as rough and smooth based on the appearance. A gulmohar is a fine textured tree when in full leaf, whereas *Spathodea companulata* is a coarse textured tree.
- The placement of all these various textures with harmony and contrast has to be achieved to get the ultimate desirable effect.

10. Light and Time

- In a garden the time factor is very important. The garden design should be planned in such a way that in the afternoon it is possible to sit in a shaded place from where the best part of the garden should be visible.
- The growth habits of the plants play a vital role in choosing the right place for them in the garden and according the layout has to be planned.

11. Tone and Colour

- In a landscape garden, the permanent backdrop is the green tones of the various trees and shrubs.
- It is possible to lay out a garden with subtle tone of entirely white or yellow flowers, but at the same time making it charming also. Another important point is that it is better to have masses of a single colour against a mixture of colours.
- A bed of roses containing only a single colour of say red, yellow, or pink has a much softer tone and beauty than a bed containing a mixture of colours. A good garden architect should have the knowledge of colour wheel and colour schemes for charming colour in the garden.

12. Mobility

- In a temperate zones, the garden changes colour very sharply and contrastingly from one season to the other thus symbolizing mobility or movement. As for example, many trees in the temperate zones attire themselves with wonderful hues due to the changes in their leaf colour in the autumn.

- In most parts of Tropical India, though these contrasting changes cannot be achieved, it is possible to bring in some subtle changes.
- For example, to create some symbol of movement trees such as Bengal or Indian Almond (*Terminalia catappa*) which changes its leaf colour into striking red twice annually before falling or *Lagerstroemia flos-reginae* which also changes the colour of the leaves to coppery shade in the autumn before shedding, or *Madhuca indica* and *Ficus religiosa*, the new foliage of these appearing as coppery red in the spring, should be planted in some parts of the garden. This, in addition, improves the landscape.
- The mobility can also be achieved by raising flowering annuals of different colour in form of flower beds. Changing the position of potted plants may also bring mobility in the garden.
- The movement and cluttering of birds will bring life and mobility to the garden. Large trees and bird-baths attract birds. For the smaller birds, the safety of shrubberies is needed to protect them from large predator birds.
- Some plants, bearing berries, such as *Ficus infectoria* and *Syzygium cumini* (Syn. *Eugenia jambolana*), can also be planted in some remote corners though they may not look very ornamental. Flowering trees such as *Bombax malabaricum* (silk cotton) or *Erythrina* also attract birds when in bloom.
- The seasonal flowers will bring in the motion and movement of colourful butterflies. Fountains or even a lawn sprinkler and streams in a garden also serve the objective of movement. The lily pools should be filled with coloured fish, the movement of which will be an added attraction.

13. Style

- Lastly, one has to decide about the style to be adopted for a particular garden. Every garden designer has to invent his own style of gardening commensurate with his budget, taste and the nature of the site, ease of maintenance.

- One can develop his own design only when he studies carefully all the garden styles of the world and grasps the underlying principles in them.

Focal Point

Mass Effect of Topiary

Space

Reterences

1. Bose, T.K., D. Mukherjee, 1977. Gardening in India
2. Randhawa, G.S. and A. Mukhopadhyaya, 1986. Floriculture in India
3. Arora, J.S. 1998. Introductory ornamental Horticulture
4. Bhattacharjee, 2004. Landscape Gardening and Design with plants.

Chapter 2

Elements of Garden Landscape and Design

There are three categories of garden elements:

1. **Major elements:** Plants and water
2. **Minor elements:** Stones, bricks, tiles, tar, metal, glass, plastic, wood, sculpture etc.,
3. **Other elements:** Light, sound, aroma, touch, food, amusement items etc.

Major Elements:

I. Plants

Classification of plants based on utilities and their functional value:-

1. **Aesthetic purpose:**
   a) Avenue planting-flower parade
   b) Ground planting
   c) Shrubbery, rockery, topiary, hedges, edges
   d) Potted plant, flower beds, boarders
   e) Ground cover
   f) Water garden
2. **Functional purpose:**
   a) Control pollution
   b) Reduces noise
   c) Control soil erosion
   d) Wind break
   e) Deciduous plants can utilize to increase temperature during winter

Criteria for selection of plants for gardening:

Certain morphological character should be considered while selecting the plants;

- Height of the plant
- Types of branches- upright, drooping, horizontal
- Spread and width of the plant
- Form of the plant

- Type of plant- evergreen, deciduous
- Colour of the plant
- Texture of the plant- smooth, rough
- Flowering time of the plant

II. Water

Water has the ability to change form, at low temperature water freezes, evaporates at high temperature and liquefies at moderate temperature.

1. **Aesthetic use**
   a) Water falls- cascade type, fountains, streams, lakes, ponds, nappe, chadar, chute.
   b) Water is used for creating reflection during night time along with lighting.
   c) The use of water helps in reducing the temperature in the microclimate.
   d) Sound of water makes a person feel comfortable.
   e) Water may mask the over power unwanted sounds
2. **Functional use** - Water will nourish the plant, increases growth and their development

Minor elements:

There are different types of minor elements which can be used in garden landscape are as follows:

1. **Stones:**
    - They can be use for creating rockery, statues or sculptures, to imitate natural water falls, garden benches, pathways, walks etc.
    - Keep in mind that stone radiates heat and will have to be used carefully
2. **Bricks:**
    - For creating garden walls, ponds
    - As paving material
    - For constructing plant boxes etc.
3. **Wood:**
    - Wood adds colour and texture to the garden
    - Used for creating paths, steps, garden bridges, ornamental picket fence, country fences, arches, pergolas etc.
    - Wood should be coated with paints/ preservatives to prevent decay
4. **Metal:**
    - For creating artistic features like garden bridges, arches, pergolas, arbour, fence, light stands, metal gardens
    - Used as a base material for plants to grow on, when used for topiary
    - Metals should be treated with anti corrosive materials before using in the out doors
5. **Sculptures**:
    - Artistic material adds beauty to the garden with the use of stones, granites, sandstone, marbles, metals.
    - Sculptures includes birds, human being, animals etc
    - It can be kept in front, middle part, and in running streams of the garden

Stones

Metal

Tar

Concrete

Bricks

Metal

6. **Glass:** It includes mist chamber, green house, conservatory, terrariums and illumination purpose
7. **Concrete**: Used for creating drives, paths, walks, ground cover
8. **Tar/ Asphalt:** Creating drives
9. **Plastic:** Drippers, sprinklers, pots, waste bin etc

Other elements:

1. **Lighting:** Illumination, focusing the focal points, illuminating water
2. **Sound:** Waterfalls, running streams, musical fountain, and wind chimes
3. **Aroma:** planting of aromatic plants in different parts of the garden, trees, aromatic plants, annual beds can be used.
4. **Touch:** Texture of materials like plant surface, sculptures, paving.
5. **Food:** Created at outside boundary of the gardens

Basic Pattern in Garden Design

1. **Circular pattern**- series of circles can utilize to create circular. It is used in formal and informal gardens.
2. **Diagonal pattern-** draws a grid line at 450 to the boundary. It is also used in formal and informal gardens.
3. **Rectangular pattern-** it is utilized in formal garden in a symmetrical manner.

Steps in Garden Design

Identification of site: Identify the site for two purposes:

- Public garden: give importance for likings of the public
- Private garden: Give importance for owners comfort

Analysis: Need to study two factors:

- Physical factor- climate, weather, soil type, existing vegetation, existing manholes, roads, path
- Social factors

Measuring up

- Draw rough sketch of the area like existing features, length and width of all features

Drawing to the scale

- Draw rough sketch of the area like existing features, length and width of all features
- Based on sheet available for small garden 1: 50 and for big garden 1: 100

Evolving a pattern

- Fence, lawn, flower beds, hedge, edge, border, waterfalls, rock garden, pond, shrubbery, island beds, carpet beds, standards etc

Turning pattern in to reality

- Mark the area using pegs and bars and plant the permanent features on marked ground

Xeriscaping

- This is a technique use to practice water conservation in creative landscape, this can be practiced in areas where scarcity of water. This can be achieved by:
    - Grouping/election of drought tolerant plant, trees, shrubs, annuals, cacti and succulents
    - Mulching
    - Ground covers
    - Minimize water loss by providing drips

Reterences

1. Bose, T.K., D. Mukherjee, 1977. Gardening in India
2. Randhawa, G.S. and A. Mukhopadhyaya, 1986. Floriculture in India
3. Arora, J.S. 1998. Introductory ornamental Horticulture

Chapter **3**

Propagation Techniques of Ornamental Plants

There are different methods of plant propagation, which can be classified under two heads:-

- Plants have the natural mechanism to multiply themselves by seeds and vegetative parts.
- The gardener has to learn the different methods of propagation to replace the old or the diseased stocks and to meet the increasing demand of a growing garden.

Methods of propagation:

1. Sexual propagation
2. Asexual or vegetative propagation

Sexual propagation

Plants are propagated sexually by seeds and spores. A seedling may not be an exact replica of the parents as a result of natural cross-fertilization or segregation of characters. Rising of plants through seeds is necessary to obtain new plants or hybrids. Seed propagated plants have long juvenile phase and takes more time to come into bearing.

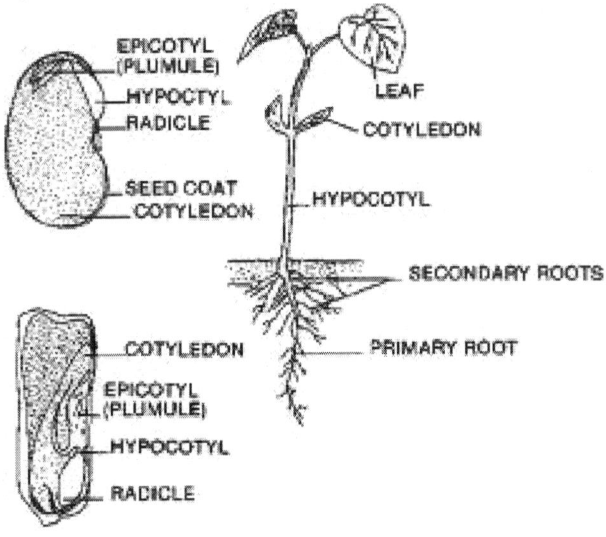

Factors controlling germination of seeds:

- When all conditions are normal, most of the seeds will germinate within a short period of time, but some may fail to do so because of immature embryos, dormancy and hormonal imbalances etc. Normally, an embryo has sufficient food material stored for germination, but in such isolated cases as the orchids, nutrients have to be provided externally to ensure germination.
- The factors such as age, stage of, maturity and viability of seeds are interrelated. Viability of seeds determines their power to germinate. The poor viability of a seed may be due to the fact that it was harvested before full maturity or it might have been harvested from a weak plant and from plants grown under improper humidity and temperature.
- Viability also greatly depends upon the age of the seeds. Most of the annuals lose viability within 6-12 months. Water is necessary since plant food should be made available to the embryo in a dissolved state. Moreover, water also softens the hard seed coat and encourages imbibitions and ultimately early seed germination.
- A quantity of water enough to completely saturate the seed coat should be used as too much watering is not good for germination. Excess of water may block the supply of oxygen or help the growth of disease-causing organism.
- On the other hand, some of the xerophytes and annuals such as *Portulaca* and *Eschscholtzia* need less qualities of moisture for germination, whereas seeds of aquatics such as water lilies and lotus remain submerged.
- The temperature should not be too high or too low. Some of the annuals such as Nemophila and Larkspur germinate better when the nights are really cool.
- Light is not essential for germination for most of the species. Seeds of some species are indifferent to light, while others are slightly inhibited (e.g., larkspur); some are completely or partially inhibited (e.g., *Nigella*

arvensis); still some others require light and geminate without cover (e.g., members of family Gesneriaceae, Ranuncululs, and Veronica).

Delayed germination and pre-treatment of seeds: Some seeds do not geminate easily for a variety of reasons, the most common causes being dormancy, rest period, and presence of hard seed coat.

Pre-treatment methods

1. **Scarification**: Seeds with hard coats require some kind of external treatments for germination. The methods commonly employed are cracking of the coats by mechanical means, abrasion, soaking in water or acid, and stratification.
2. **Stratification**: consists of fully exposing the seeds to the action of the weather by placing them in alternate layers of moist sand or sifted soil either in shallow boxes or in flats or in trenches outdoors. The flats and boxes are placed outdoors to be exposed to rain and snow and to freeze and thaw. Rose seeds are thus treated to get better germination. Larkspur also germinates better during the summer if stored in a cool moist place for several days before sowing. Orchid seeds are so minute in size that they have very little or no stored food to nourish the growing embryo. Hence these are germinated in association with VAM fungi which provides the nutrition during the initial period of germination process.

Procedure for sowing of seeds

1. The seed pans of wooden or plastic materials are generally used for raising seedling of ornamental plants.
2. The standard dimensions of the pans are 100x40 cm and may vary sometimes.
3. Sterilization of media is done either by moist steam or solarization.
4. The seed pans are filled with sterilized germination media. Fine sieved sand mixed with coco peat is found best for the germination.
5. The media is compacted by even pressing it with wood hammer.
6. The sowing lines or grooves are to be made with the help of stick of pencil thickness at 5 cm apart from each other. The depth of the groove attained will be 0.5-1.0 cm.
7. The seeds are sown into the grooves made. If the seeds are bold in size then cover them with fine sieved sand however, there is no need to cover the seed if those are tiny in size.
8. A spray of water mixed with fungicide is recommended immediately after the sowing.
9. The pans are covered with a polyethylene sheet and then a sheet of news paper above. This is done to create congenial environment for germination.
10. The emergence of radical can be observed in one or two days which is the indication to remove the covering material.

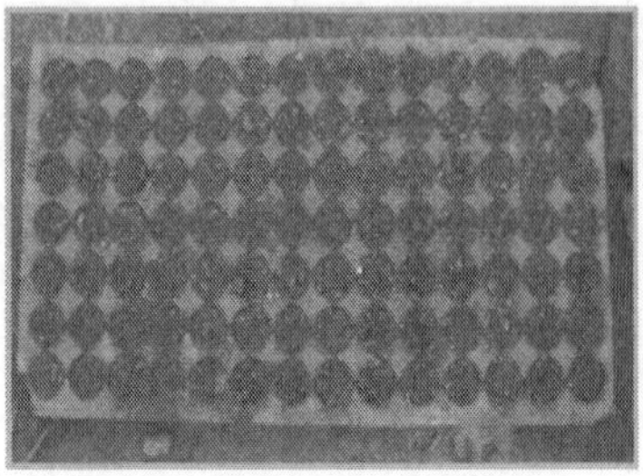

Pro-trays for Seed Sowing

Seed of *celosia* (Cocks Comb)

Germinated Seedlings

Spores: Actually, a spore (as in ferns) is an asexual body. But, when this falls on a moist surface it produces small plant bodies (prothallia), which develops the sex organs (archegonia and antheridia) and these in due course develop sex elements (gametes), which fuse and resulting body develops into the fern plant.

Asexual propagation or Vegetative Propagation

The propagation of plants from vegetative plant parts such as buds, cuttings, roots etc is known as vegetative propagation. There are different methods of vegetative or asexual propagation such as cutting, layering, division, separation, budding, grafting etc. excluding tissue culture.

1. Cuttings

1. It is defined as a process by which a plant is produced by severing a vegetative portion from the plant and rooting it in a favorable medium under optimum conditions.
2. The plant parts that are used for this purpose are stems, leaves, and modified stems such as tubers, corms, rhizomes, runners, and bulbs. In general, propagation by cutting is the cheapest and the most convenient method and hence this is used more popularly to raise new plants.

There are different types of cuttings

a) Stem cuttings: Stem cuttings are of four types-

1. Herbaceous Cuttings
2. Soft Wood
3. Semi-Hard Wood Cuttings
4. Hard Wood Cuttings

Hardwood cutting

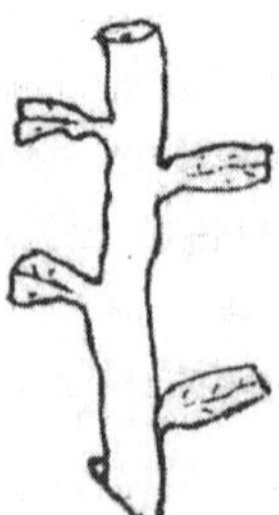

Semi hardwood cutting

softwood cutting

Note: Cuttings with leaves are preferable over leafless cuttings.

1. **Herbaceous Cuttings**

   This type of cuttings is taken from succulent herbaceous green house plants.

   Preparation and planting: Select the succulent herbaceous shoots from a healthy and vigorous green house growing plant. Retain all the leaves. Give a basal cut below a basal node. Plant the cuttings in the same way as hard wood cuttings are planted. *Eg.* Chrysanthemum,Coleus, Carnations, Geraniums, Cactus etc.

2. **Soft Wood Cuttings**

   - Soft wood cuttings may again be separated into two distinct categories:
   - Those taken from herbaceous plants such as Carnation, Chrysanthemum, Coleus, Dahlias, Delphiniums, Petunia etc. and
   - The un ripened tips of woody plants such as most of the ornamental shrubs and some trees.
   - Soft wood cuttings are taken from below a node and the bottom leaves are removed before planting.
   - The soft wood or herbaceous cuttings are generally 2.5 cm to 10 cm long.

3. **Semi-Hard Wood Cuttings**

   - The portions of the stems which have passed the soft wood stage but are not yet mature are preferred to as semi-hard wood cuttings.
   - Semi-hard wood cuttings are used and the bottom leaves are removed.

4. **Hard Wood Cuttings**

   - These are cuttings of shrubs and trees taken from mature current years growth.
   - The length of cuttings varies with the type of plants and weather conditions.
   - One meter to one and a half meters long hard wood cuttings in some trees and shrubs. Examples: *Ficus benghalensis*, Citharexylon, *Gliricidia maculate*, and species of Erythriana, when planted out in the open, root easily.
   - Generally hard wood cuttings are 15-30 cm long.

Hard wood Stem cuttings are of three kinds:

a) **Terminal Cuttings or Straight cuttings**: The lateral shoots which form the tip portion of the shoots. The lower leaves are removed by snipping (not tearing off) and only about four terminal leaves are left in such cuttings.

b) **Heel Cuttings**: The lateral shoots which re pulled off from the stem and contain a portion of the stem are known as heel cuttings; these root more successfully than terminal cuttings, the rough wood and bark at the bottom may be smoothened with a sharp knife.

c) **Node cuttings**: Single or multiple node stem cuttings are also obtained from plants such as Dieffenbachia and Dracaena and these are inserted horizontally in sand.

- With the present-day facilities such as mist propagation, propagation frames, bottom heat, etc., it is possible to raise cuttings any time of the year. But it has been found that the best time for taking cuttings corresponds to the time of the growing season of the parent plant.
- In most parts of India, the duration of this season is from July to September.

Soft-Wood Cuttings

Semi-Hard-Wood Cuttings

Hard-Wood Cuttings

Tip Cuttings

Node Cuttings

b) Leaf Cuttings

Certain plants with thick and fleshy leaves have the capacity to produce plantlets on their leaves. In leaf cuttings, the leaf blade with or without petiole and axillary bud is used for starting new plants. Adventious roots and shoots form at the base of the leaf and form in to a new plant. However, the original leaf does not become a part of the new plant. Frequent watering and high humidity and bottom heating are desirable for better and rapid rooting of leaf cuttings. Sand or sand and peat moss (1:1) are satisfactory rooting media for leaf cuttings. For leaf cuttings, depending on the species the whole leaf blade, leaf blade sections or the leaf with petiole is used.

So, leaf cuttings can be classified in to:

1. Leaf blade cutting
2. Leaf vein cutting / Leaf slashing
3. Leaf margin cutting
4. Leaf bud cutting

Leaf blade/Leaf section cutting: ***Preparation and planting:*** Select a healthy leaf and Give a slanting cut towards the base of the leaf. Measure a length of about 7 to 10-cm and give a horizontal cut towards the terminal end. Prepare as many cuttings as possible from the selected leaf. Insert up to ¾ of the prepared leaf cuttings in to the medium. Take care of polarity while planting the cuttings. Compress the soil around the leaf cuttings and water immediately. Eg. *Sansevieria.*

Leaf vein cutting/Leaf slashing: ***Preparation and planting:*** Select a healthy and full mature leaf and detach it from the mother plant. Give cuts to alternate veins closer to the petiole on the lower surface of the leaf. Keep the leaf flat on the medium in such a way that the lower portion comes in contact with the medium. Pin or hold down the leaf in some manner so as to expose the upper surface and to maintain the contact between the cuts on the vein and the rooting medium. Water the cuttings carefully Eg. *Begonia rex.*

Leaf margin cutting: ***Preparation and planting:*** Select a mature and healthy leaf with the foliar embryos intact. Keep the leaf flat on the rooting medium. If the leaf is folded, just cut along the mid rib, so that the leaf can be kept flat on the medium. Keep some weight on the leaf or partially cover it with soil, so that the margin comes in contact with the medium. Water the cuttings carefully Eg. *Bryophyllum.*

Leaf bud cuttings: This cutting consists of a leaf blade, petiole and a short piece of the stem with attached axillary bud. This is practiced in species that are able to initiate roots but not shoots from the detached leaves. In such case the axillary bud at the base of the petiole provides for the essential shoot formation.

Preparation and planting: Select a healthy and mature shoot with well developed buds and healthy active growing leaves. Separate each leaf along with the axillary bud and a small portion of the stem. Repeat the process until possible number of leaf bud cuttings are made. Treat if necessary the cut surface of the prepared cuttings with the recommended root promoting substance to stimulate rapid root formation. Insert the prepared cutting in the rooting medium so that the bud is1.5 to 2.5 cm below the surface. Compress the medium around the cutting and water immediately. Eg., Camellia, Rhododendron etc.

Leaf Section cutting—Sansevieria

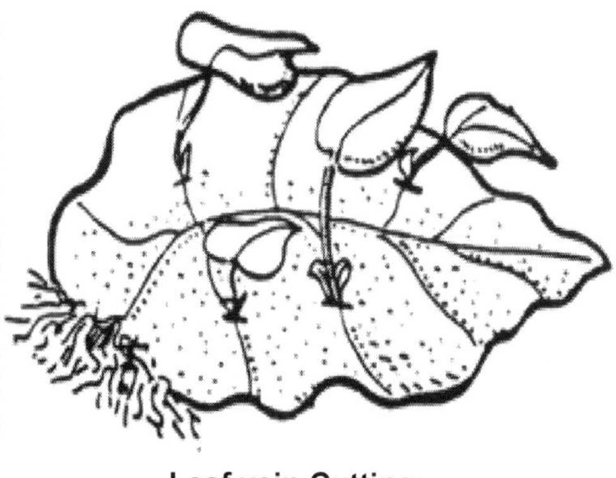

Leaf vein Cutting

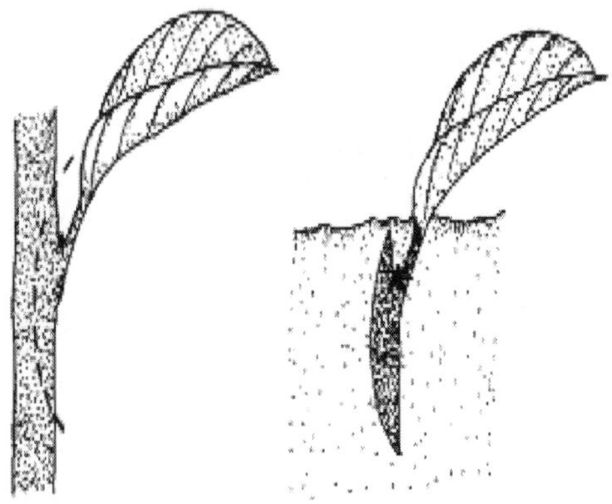

Leaf bud cutting

cutting Leaf Margin Cutting of Bryophyllum

c) Root cuttings: Plants which give rise root suckers freely are propagated by root cuttings.

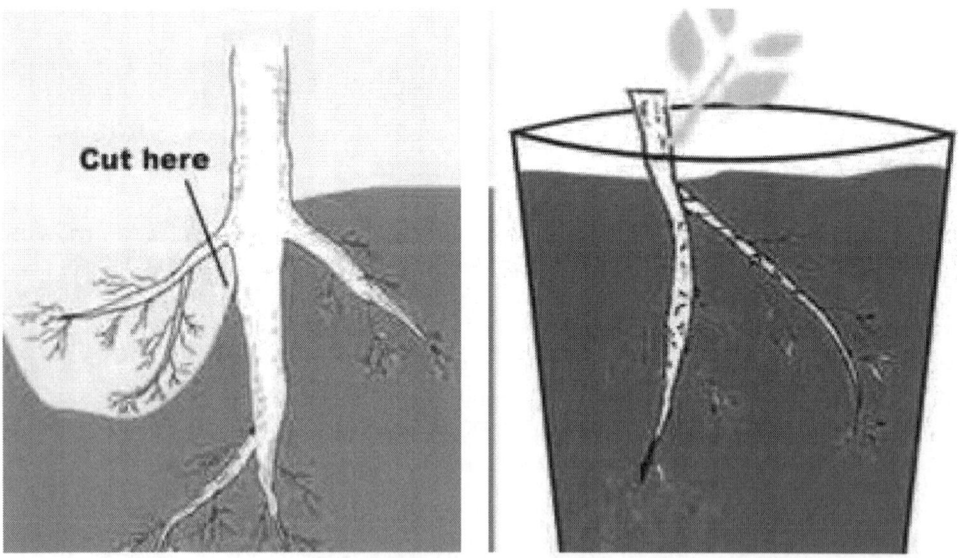

Root cuttings

Plant propagation by Layering

Layering is the developing of roots on a stem while it is still attached to the parent plant. The rooted stem is then detached or become a new plant growing on its own roots. A layered stem is known as a layer.

Advantages of Layering

- It is an easy method and does not require much care and arrangements like cuttings.

- The mother plant supplies nutrient and other metabolites as it remains attached while rooting.
- Some of the species which do not root by cuttings can be propagated by layering

Disadvantages

- A number of new plants that can be produced from any given number of plants or mother plants by layering is low when compared with cutting
- It is very costly method where labour charges are very high

  Methods of layering can be broadly divided into two main classes:-

  - Ground layering
  - Air layering

A. Ground layering

- In these methods, the rooting of layers takes place in the ground media or in pots containing rooting media.

  Different kinds of ground layering are as follows:-

  1. **Simple Layering**: It consists of bending down a shoot and burying a part of it in the soil so that the tip is in air. The buried portion is wounded by a single stroke upwards on the underside of the shoot to be buried. The layered branch is held firmly in position by pegs and large stones on the surface of the soil above the bending. Roots are formed at the buried portion where the cut has been made. The bent portion is completely covered with soil. Eg. Rose, Jasmine, etc.
  2. **Compound or Serpentine Layering**: This is similar to simple layering except that the flexible branches are covered with soil in several places. The shoots are alternately covered and exposed over their entire length. Eg. Clematis, Peperomia etc.
  3. **Trench Layering or Etiolation Method**: Etiolation refers to growing in darkness and in this method; a branch is covered over its entire length to produce from all nodes or buds, wherein roots are formed. One year old plant is planted slantingly at an angle of 30º to 45º and about 1m apart in the ground. Once these plants are established, they are bent and laid flat over the bottom of a trench of about 5 cm deep. When buds start to swell, a layer of 2.5 cm soil is placed on the branch. As the shoots grow, more soil is added until they are covered to a depth of about 12-15 cm so that the shoots are etiolated to form roots.
  4. **Mound Layering or Stooling**: It is a form of layering, in which the already established parent plant is cut back to the ground and resulting growths from the stud or stool are covered with soil to one-half its height. The soil is applied at intervals as the shoots grow.
  5. **Tip layering**: The tips of current season's shoots are buried in the soil. The tip of the shoot curves upward to produce a sharp bend in the stem from which roots develop.

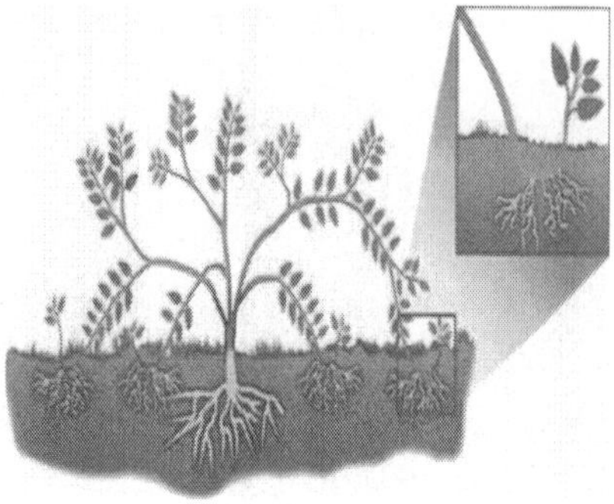

Ground Layering

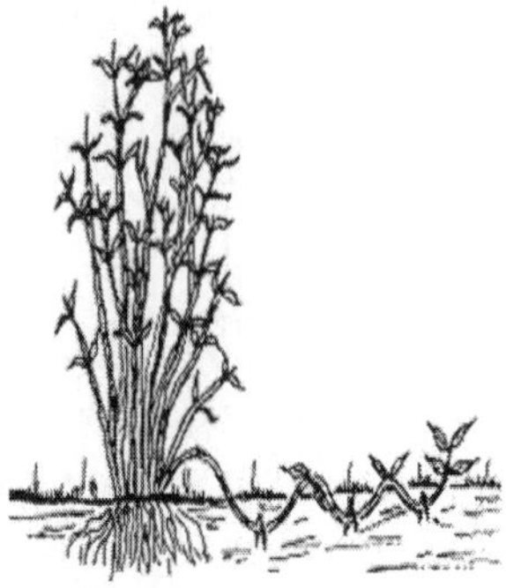

Serpentine/Contineous Layering

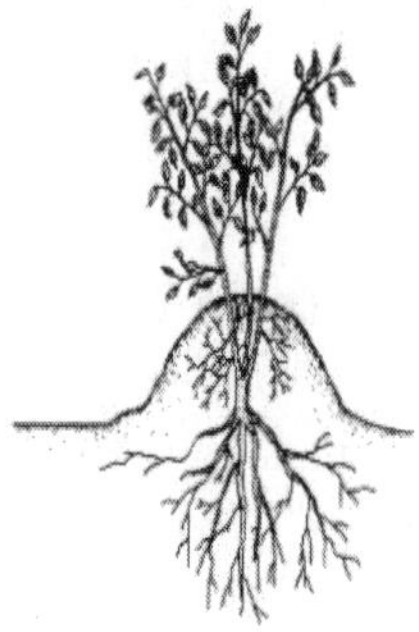

Trench Layering

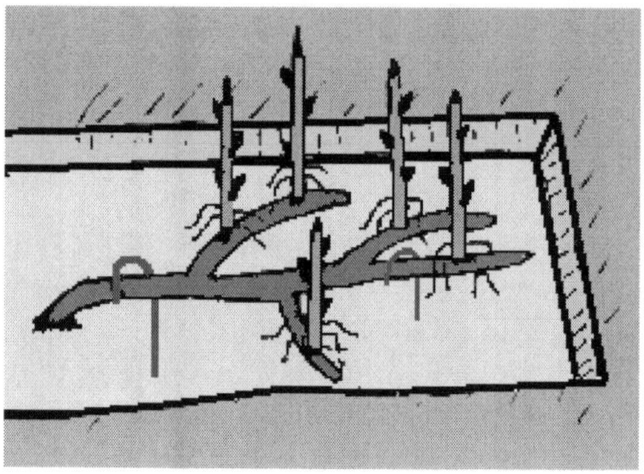

Trench Layering

B. Air layering

- This method is also known as Chinese layering, pot layerage, gootee, marcottage or cir-cumposition.
- In this method, the roots are induced to form on the aerial part of the plant where the stem is girdled or slit at an upward angle.
- Girdling consists of removal of a strip of bark of 2.5 to 3 cm wide around the stem.
- Scraping the exposed surface to ensure complete removal of phloem and cambium is desirable to retard healing.
- The injured portion is enclosed with rooting medium like Sphagnum moss or Vermiculite which is maintained continuously moist by wrapping with polythene sheets.
- The polythene sheets permit gaseous exchange but are impervious to water.
- The roots are formed on the upper portion of the cut end of the ring. When the stem or shoot has produced a good root system, a first cut is given half way through the stem just below the point of rooting. Next cut is given 15 days later and the rooted stem removed.

- As the root system is small when compared to the shoot system, the rooted layer is planted in pots to produce more roots before planting in the field.

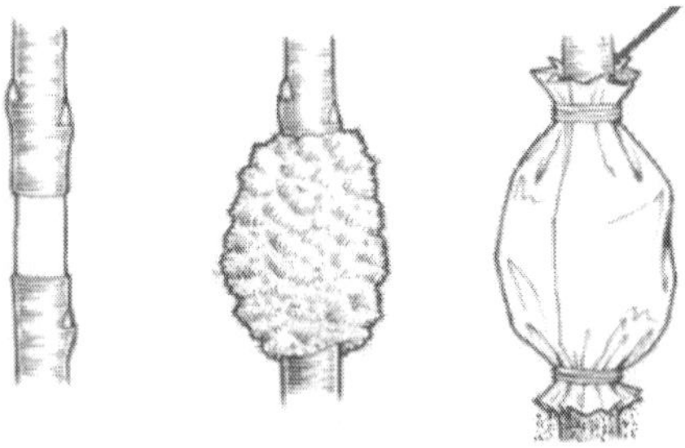

Air Layering

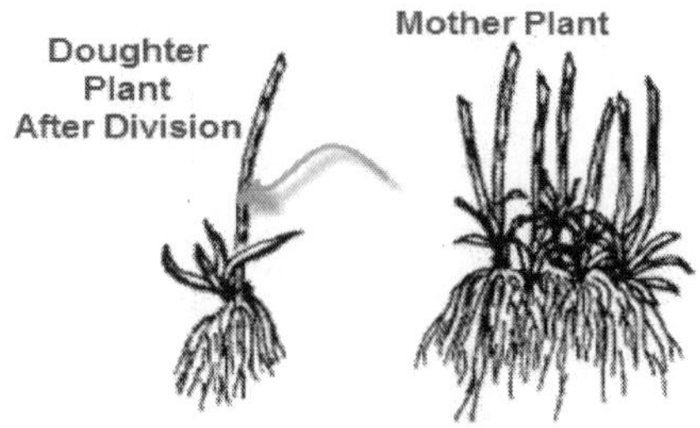

Division

Divisions

- Shrubs may be divided in the same manner but when the plants are dormant. Because shrubs are often woody, they may need to be separated with a shovel or hatchet. Before planting, trim back the shoots and cut off damaged roots.
- Shrubs that may be divided include red osier dogwood, summersweet clethra, deutzia, euonymus, hydrangea, kerria, spirea and snowberry. Herbaceous plants such as Chrysanthemum, Day lily, Peony and Iris can also be divided by this method.

Plant Propagation by Budding

This is a grafting technique in which a single bud from the desired scion is used rather than an entire scion containing many buds. It is an art of inserting a bud on the root stock in such a way that both will unite and continue to grow as a single individual plant.

Types of budding

1. **Shield or 'T' budding or '?' budding**
   - In the root stock, a transverse or horizontal cut of 1 to 1.5cm length is made first. Below or above this cut a vertical cut of 2.5 to 3 cm length is made and connected to the horizontal cut.
   - Two flaps of the bark should be opened with help of knife (ivory edge).
   - The cuts are given in the stock at the height of 5 to 25cm above the soil in a smooth bark surface.
   - In the scion, 1.25cm above the bud, a slanting cut is made and 2.5cm long bud is taken in the shape of a shield. Insert the bud by pushing it downward under the two flaps of bark (Horizontal cut of stock and scion should be even).
   - Then budded portion is covered with plastic tape or adhesive tape.
2. **Patch budding**:
   - It is called so because a patch of scion and root stock is used in this method. In this method a rectangular patch of bark is completely

removed from the stock plant and is replaced with a patch of bark of the same size containing a bud from the desired mother plant.

- For this method to be successful the bark of stock and bud stick should be easily slipping. The diameter of the stock and bud stick should preferably be about the same.

3. **Chip budding**

- It is being practiced at time when bark is not slipping from mother plant.
- In this method bark with some wood is removed from but stick as scion and used for budding. This is the only method of budding that can be done even when the bark is not slipping easily.
- This method can be used with fairly small material 1 to 2.5 cm in diameter.
- A chip of bark and wood is removed from a smooth place between nodes near the base of the stock and replaced by another chip of the same size and shape removed from the bud stick which contains a bud of the desire cultivar.

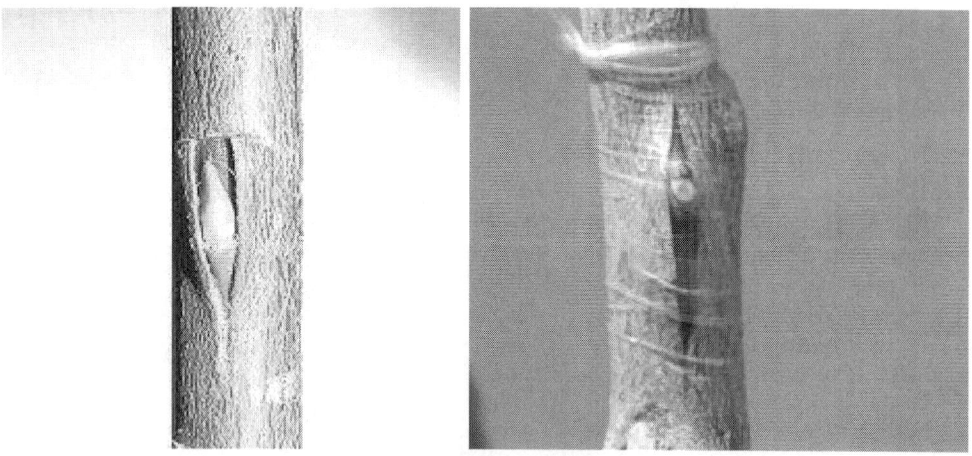

Shield budding / T- budding

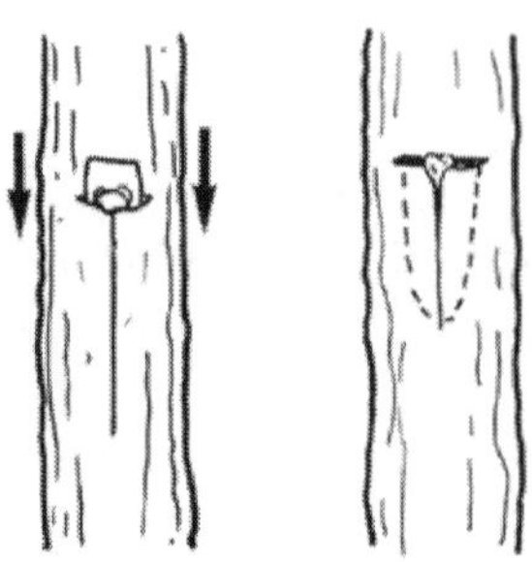

T-Budding/Shield Budding

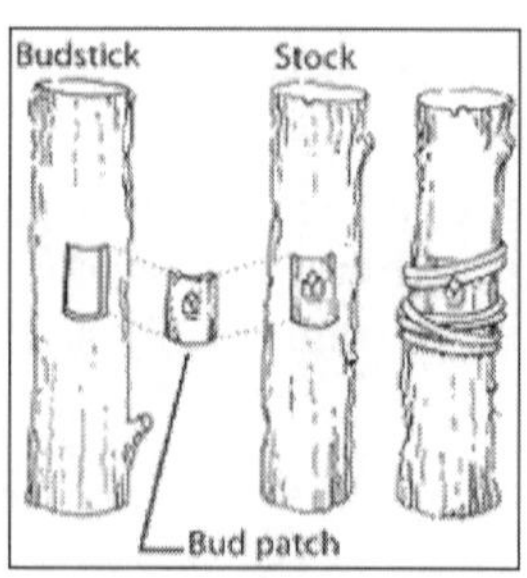

Patch Budding

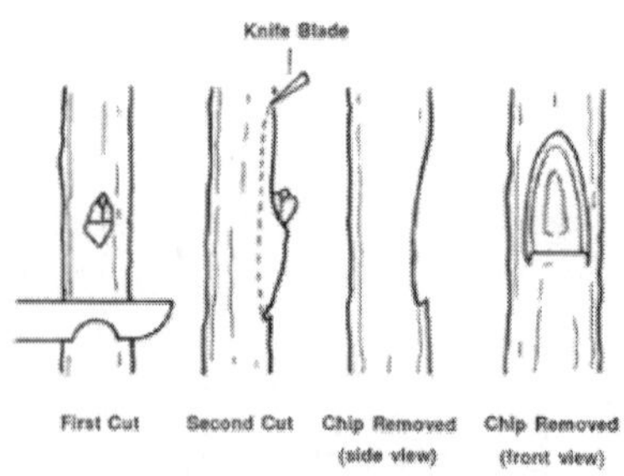

Chip Budding

3. Ring Bidding

The bud is prepared by taking a ring of a bark, 3cm long with the bud in the centre. In the root stock, two transverse cut 1.5cm apart are made and these are connected with a vertical cut and a ring of bark is removed. The prepared scion bud with the ring of bark is fitted in the exposed portion of the rootstock and tied. Eg., Cinchona.

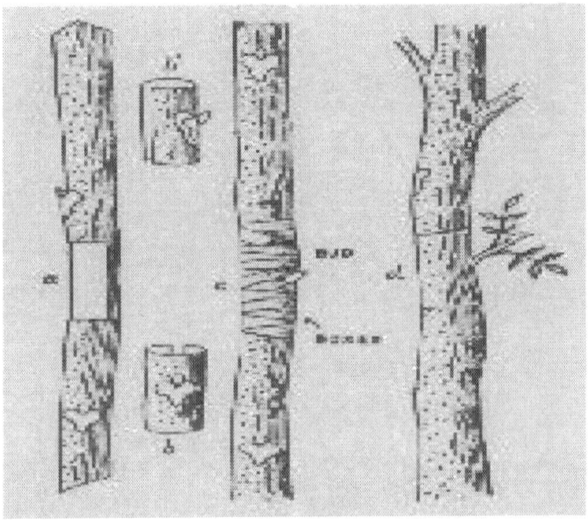

Ring Budding

Plant Propagation by Grafting

Grafting is an art of inserting a part of one plant into another plant by exposing the actively growing tissue so that they will unite and continue their growth as one plant. **Scion** is the upper part of the graft and from which stem and branches will grow into a plant. **Root stock** is the lower part of the graft and this forms the root system of the grafted plant. Root stock is also called as stock or under stock.

Types of grafting: Mainly in grafting there are two types:

A. Attached scion methods of grafting

B. Detached scion methods of grafting

In attached scion methods of grafting the scion is still attached to the mother plant (Scion Plant) till the graft union takes place where as in detached scion methods of grafting the scion is separated from the scion plant or mother plant just before grafting. Under attached scion methods of grafting simple inarching or approach grafting is most important.

1. **Inarching or approach grafting**: In this method rootstock are raised in pots. Then they are brought near the mother plant. Here scion remains in mother plant. One year old seedlings of pencil thickness are selected. Above ground level at 15 to 20cm height in rootstock, 5 to 8cm long slice of bark with wood is removed. This cut should be smooth and it tapers gently towards the tip and bottom. Same type of cut is made on scion and the two cuts are placed face to face and tied firmly with banana fiber and then with twine over it. After that union is covered with a mixture of cow dung and mud in equal parts. After 6-8 weeks top of rootstock is removed

above graft union and base of scion below the graft union. First half cut is given and another half cut is given after an interval of 10 days.

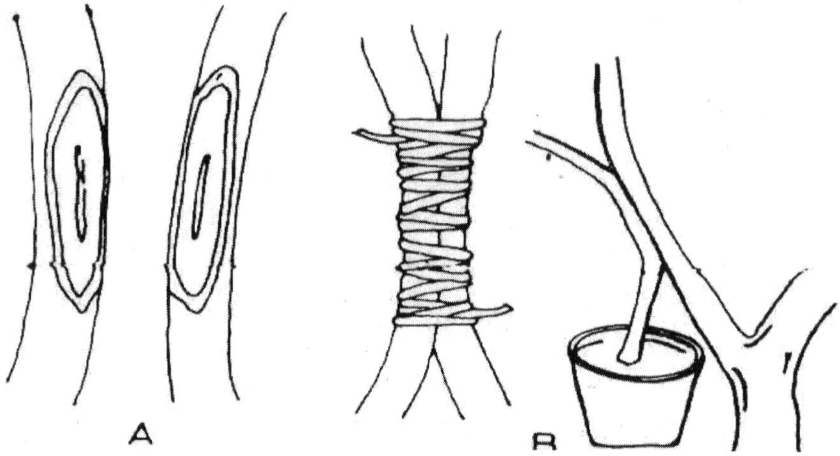

Simple Inarching

2. **Epicotyl or stone grafting**: Seeds are raised in bed and the germinated seedlings of 8 to 15 days old are taken out and grafted indoor by beheading the seedling about 5cm above the seed and then inserting the wedge shape scion in the vertical split at the beheaded stock. Polythene tapes at 200 gauge thickness are utilized for tying the graft. The grafts are planted in poly bags filled.

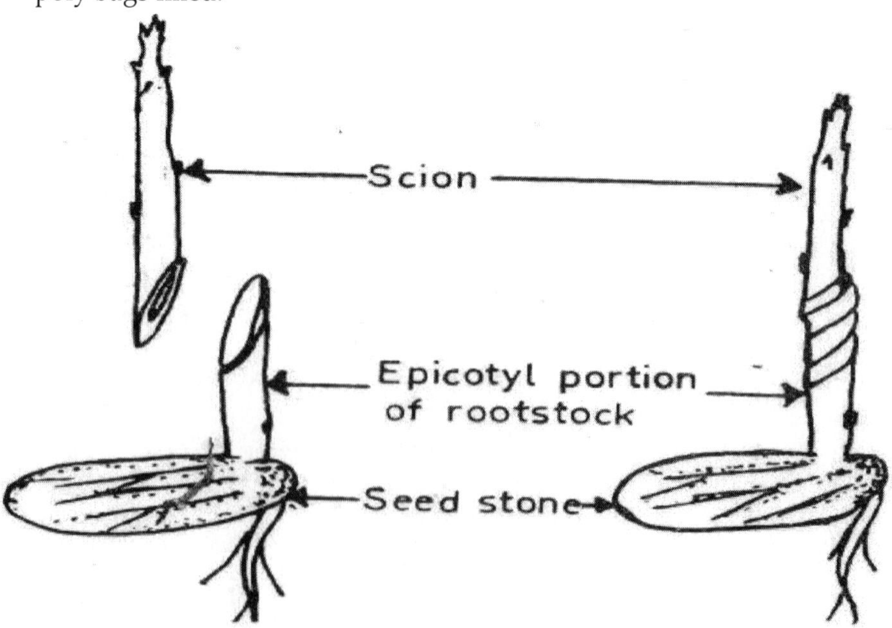

Epicotyl Grafting

3. **Cleft grafting**: The base of the scion is prepared in the form of a wedge. The rootstock is split in which the scion is inserted. This method is usually done on thick stocks of 2 – 8cm in diameter. This is common method followed in top working of trees. The stock is given a smooth cut and then it is split at the centre and two scions are inserted at the ends in such a way that the cambial layers of stock and scion are in contact. After the successful graft union, one of the scions, which is well developed is allowed to grown.
4. **Veneer grafting**: This is modification of side grafting. In this method, a shallow downward and inward cut of 2.5 – 4cm long is made in a smooth area just above the crown of the stock plant. At the base of this cut, a second short inward and downward cut is made intersecting the first cut, so as to remove a piece of wood and bark. The scion is prepared with a long cut along one side and a very short one at the base of the scion on the opposite side. The scion cuts should be the same length and width as those made in the stock so that the cambium layers can be matched as closely as possible.

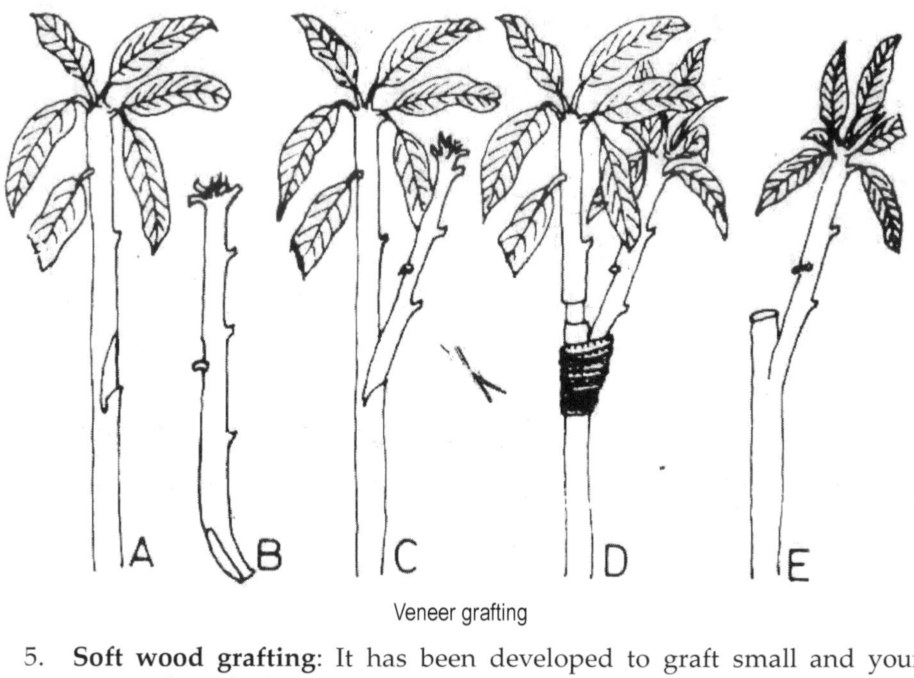

Veneer grafting

5. **Soft wood grafting**: It has been developed to graft small and young rootstocks which are grown *in situ* or in pots. Eg. Cashew, Mango, Jackfruit. ***Procedure:*** Raise the rootstock seedlings in suitable containers or preferably in the main field itself where the grafts are desired to be grown and allow them to grow for a year or more. When the seedling attain a height of 30-45cm and the new shoot and leaves usually have bronze colour. Decapitate the top portion of the fresh growth on the stock plant with a knife, retaining bout 8 cm of the fresh stem. Make a longitudinal cut of 3 cm in the retained fresh stem. Select a scion stick of about 10cm long

and about the same thickness as of the prepared stem on the stock. Cut the basal end of the scion to a wedge shape of about 3cm long by chopping the bark and a little wood on two opposite sides. Insert the prepared wedge part of the scion stick into the slit made on the stock and secure firmly with polythene strips. Water the grafted plant regularly. The scion sprouts in about three weeks.

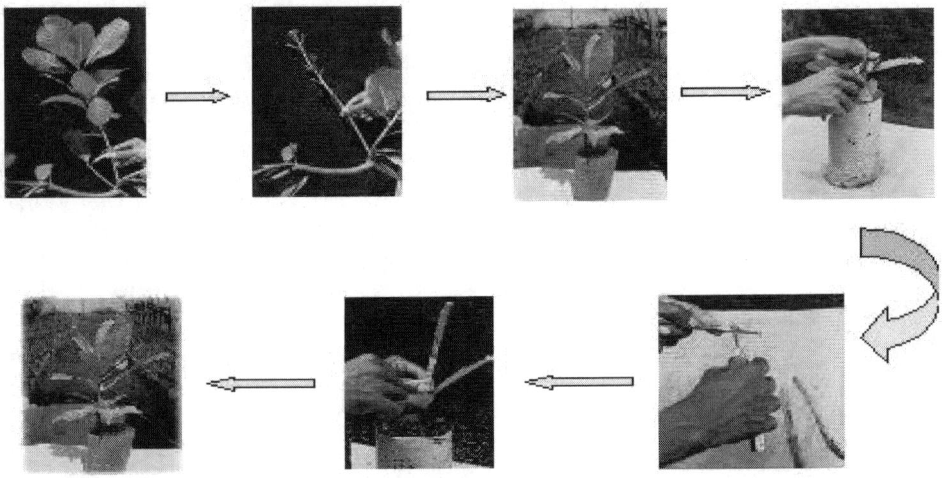

Softwood grafting

6. **Top working:** Top-working for changing a variety is generally done on long lived species, growing in a healthy condition. Short lived species, old trees or diseased trees are not suitable for top working; in such cases new planting is considered more economical and useful than top working. This practice is resorted to (1) when the existing tree is of inferior type, (2) when the tree is unproductive and (3) to provide pollenizers (4) to change the variety. For top working different methods of grafting like cleft grafting, bark grafting, splice grafting or side grafting can be used. However, cleft is the most popular and commonly used method for top working especially when thick branches are selected .When younger and thin branches are used, whip and tongue grafting are best. Top working of older trees is generally done over a period of two years. In the first year, half of the scaffold branches are top worked retaining the other branches as nurse branches which in turn are grafted in the second year. In the smaller and comparatively younger trees the entire tree is top worked in the first year. Here also one or more nurse branches are retained till the union is successful. Nurse branches protect the top worked scions from winter injury, sun burn and also from desiccating winds and water sprouts develop less frequently when nurse branches are retained. Top working is most successful when relatively young trees are used. If older trees are selected for top working, it is better to select vigorous lateral branches that arise from the main limbs. The branches to be top worked

should be cut in such a way that the cut surface is smooth and is at a point of the branch where there are no knots or smaller branches. Immediately after top working the limbs should be thoroughly covered with grafting wax, sealing all the exposed cut surfaces.

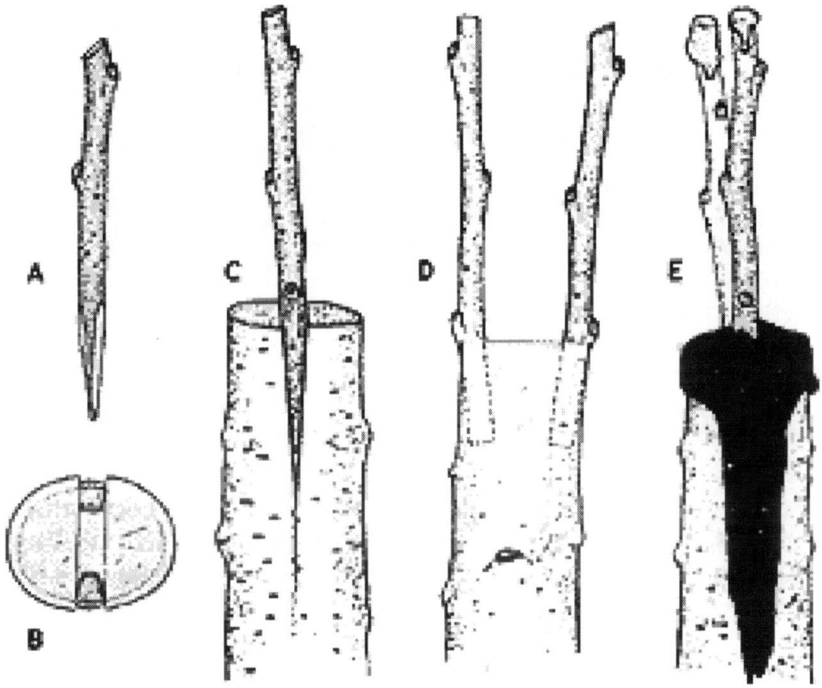

Top working by cleft grafting on main branches

References

1. Bose, T.K., D. Mukherjee, 1977. Gardening in India
2. Randhawa, G.S. and A. Mukhopadhyaya, 1986. Floriculture in India
3. Arora, J.S. 1998. Introductory Ornamental Horticulture
4. Bhattacharjee, 2004. Landscape Gardening and Design with plants.
5. Chaddha K.L. 2000. Handbook of Horticulture

Chapter **4**

Graphics and Symbols of Landscape Designing

Graphic language: The mode of communication through sketches is known as graphic language.

Drawing: The graphical representation of an object is called as drawing.

Landscape drawing

- To manufacture a component, or to develop a garden, both the shapes and sizes must be known before taking up the development.
- Therefore, the drawing of an object, which contains all the necessary information, like actual shape, accurate sizes, method of development, etc., required for the development of an object is called landscape
- Since, the landscaping of an object is drawn well before the object is constructed, the person who draws it must have a clear picture of the shape of the object in his mind. Landscaping drawing may be either representational of graphic.

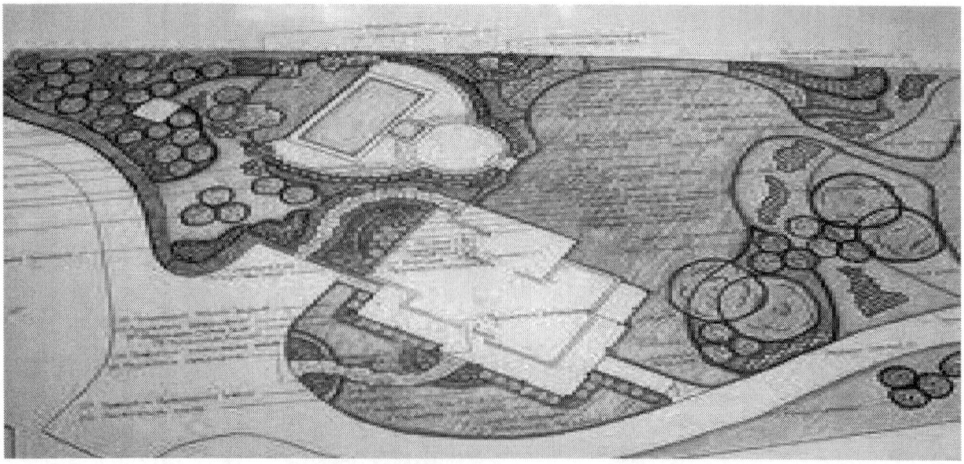

The representation drawing shows only the appearance of an object, whereas the graphic drawing shows the accurate sizes and shapes of the object. Since landscape drawing is a graphical medium of expression of technical details without the barrier of a language, it is called the **universal language of landscapers.**

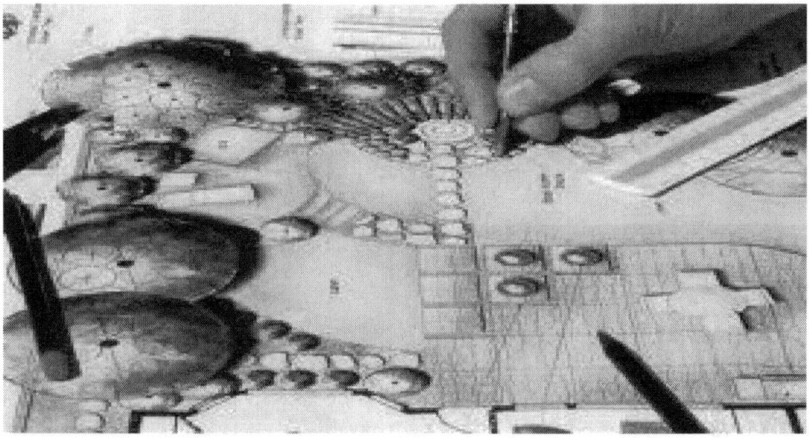

Equipment and Instruments used in landscape drawing

Since the landscape drawing shows the accurate sizes and actual shapes of the object, it must be drawn to scale with the aid of geometrical instruments-

1. **T-square:** The T-square is made of well seasoned hard wood and consists of two parts namely blade and stock. The stock slides over the working edge of the drawing board. The blade which moves on the surface of the board has a drawing edge. Its main use is to draw horizontal lines and also it serves as a support to place the set-square over it.

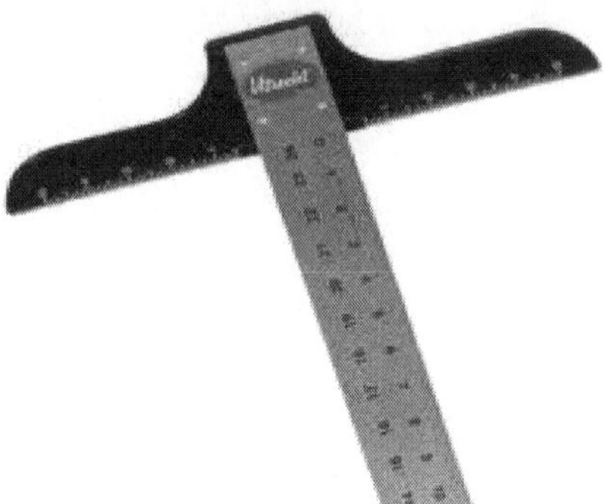

T-square

2. Drawing board

- The drawing board is made of well seasoned pine wood. This is rigid, sturdy and light in weight. Along the left edge of the board, a groove is cut, and a perfectly supporting guide for the T-square to slide on.
- The drawing board must be placed on the table with the working edge always to be at the left side.
- The T-square is made of well seasoned hard wood and consists of two parts namely blade and stock. The stock slides over the working edge of the drawing board.

- The blade which moves on the surface of the board has a drawing edge. Its main use is to draw horizontal lines and also it serves as a support to place the set-square over it.

Drawing board

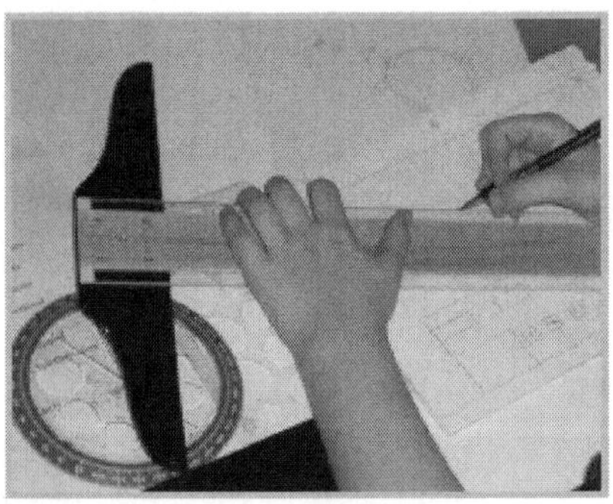

T Square

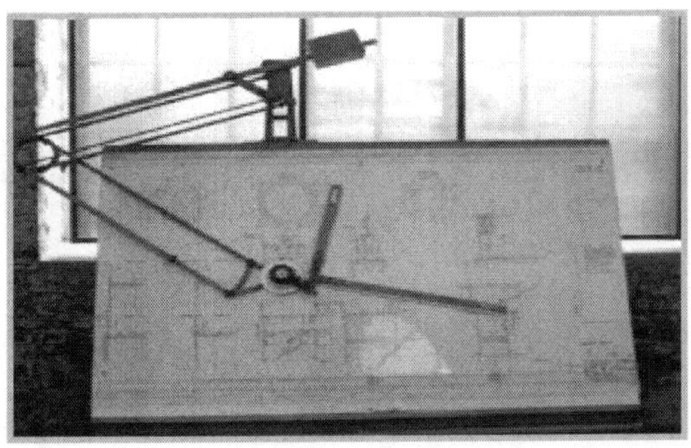

Drawing Board with Drafting Machine

3. Sets Square

- Generally, these are available in pairs. One is 60° – 30° set square and another is 45° set-square. These are used to draw vertical, inclined and parallel lines and for setting the combination of the angles in multiples of 15°.

Pencil

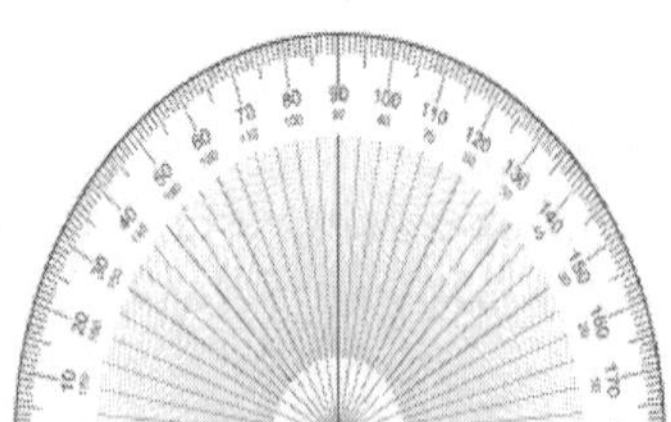

Protactor

Set Squares

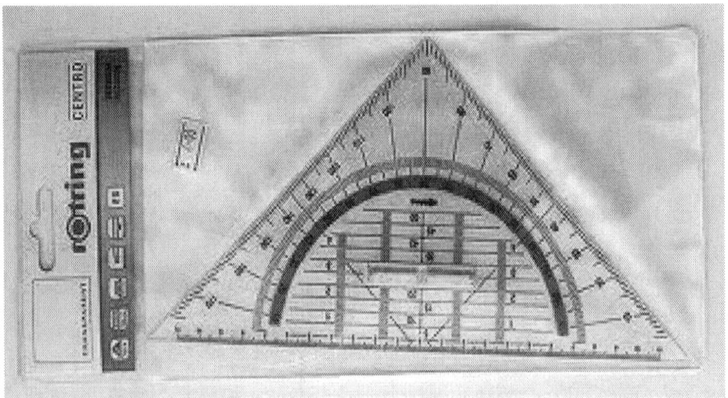

4. Protactor

- This is made of celluloid. Protractor will be either semicircular or circular. It is used to set up or measure any given angle.

5. Clinograph

- This is used to draw a series of parallel and perpendicular lines.

6. Drafting Machine and Mini Drafter

- It combines all the functions of the T-square, set-square, protractor and Clinograph. A simplified miniature version of the drafting machine called 'mini-drafter'.

- It consists of a clamp provided with a screw, two pairs of parallelogram of bars, a protractor head with a screwed knob, and two mutually perpendicular metallic or celluloid scales with ruling edges. The two pairs of parallelogram of bars are provided at right angles to a pivot place.
- One pair of parallelogram of bars is pivoted to the clamp while the other pair is pivoted to the protractor, which can be swiveled to any angular position.
- The knob provided at the protractor head facilities the clamping of the scales in any angular position.
- A drafting machine is used by professional draftsman to produce the drawings with ease and rapidity.

7. Pencils

Two types of pencil are used in drawing practice are Wooden pencils and Mechanical clutch pencils.

a) **Wooden pencil:** Wooden pencils are graded and designated by the numbers and letters from 6B, very soft and black through, 5B, 4B, 3B, 2B, B in the decreasing order of softness and blackness, and HB to F, the medium grade then H, 2H, 3H, 4H, 5H, 6H, 7H, 8H, 9H in the increasing order of hardness. For students use, H, 2H and HB pencils and finished with H or HB pencils.

b) **Mechanical clutch pencil:** Mechanical clutch pencils with 0.5mm thick lead are preferred than the wooden pencils as they need no sharpening. H, 2H and HB grade leads are used in these pencils.

8. French curves- French curves are templates of various curved shapes. The French curve is used as a guiding edge for drawing non-circular curves.

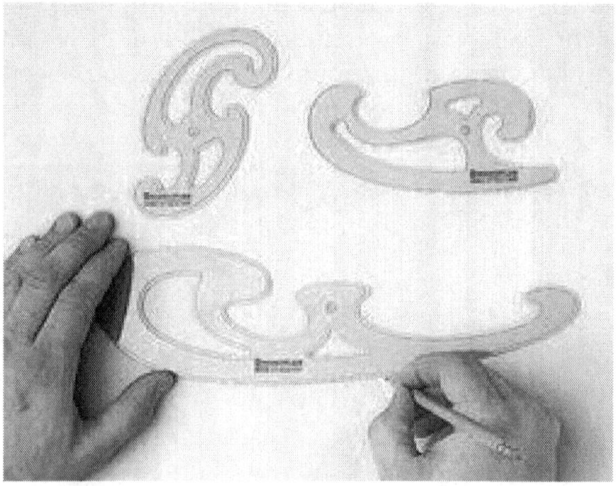

9. Drawing pins or clips or tape -These are used to secure the drawing sheets to the drawing board firmly.

10. Eraser -Soft rubber is the widely used form of eraser.

11. Erasing Shield -It is a thin plastic or metallic plate cut with slots of different lengths, widths and shapes, used to erase unwanted pencil lines without erasing the surrounding lines.

12. Drawing instrument box-The instrument box consists of a compass with interchangeable pencil or pen legs, a divider, a bow-divider and a lengthening bar.

13. Scales-Scales are the measuring devices usually made either of cardboard or plastic. They are available in a set consisting of eight scales, designated from M1 to M8.

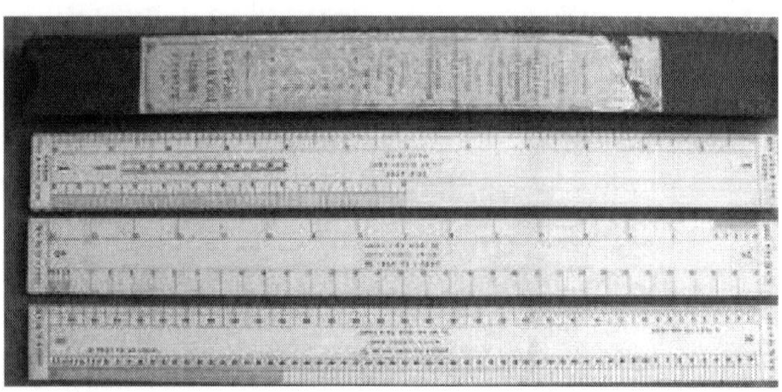

14. Drawing sheets

- The standard sizes of the drawing sheets recommended by the Bureau of Indian Standards.
- The surface area of A0 size sheet is one square meter.
- The sizes of the successive drawing sheets are obtained by halving the previous higher size sheet.

- The widths to length ratio for all sheet sizes are maintained as 1:2. A2 size drawing sheet is convenient for students use.

Layout of drawing sheets- The layout of a drawing sheet is the standard form of arrangement of the important particulars which are shown on it to ensure that all the necessary information is included in the drawing sheet and facilitate its quick reading and to make it possible for essential references to be located easily, especially when drawings are prepared by several officers.

Positioning of drawing sheets- The drawing sheets may be positioned with their longer sides being placed either horizontally or vertically depending on overall sizes of the drawing to be drawn.

Borders- The border is the space left all round in between the trimmed edges of the drawing sheet and the frame. It has been recommended that these borders have a minimum width of 20mm for A0 and A1 sizes, and a minimum width of 10mm for A2, A3 and A4 sizes.

Filing Margin- A filing margin may be provided for taking perforations. This margin shall have a minimum width of 20mm with the border included therein.

Grid Reference system- A grid reference system is provided for all the sizes of drawing sheets to facilitate easy location of any portion of the drawing within the frame. The length and width of the frames are divided into even number of divisions. The number of divisions chosen for a particular sheet depends on the complexity of the drawing.

Tile Block

- A title block is an important feature provided on a technical drawing or with an associated document for recording the technical and administrative details.
- It is placed on the right hand bottom corner of the drawing sheet.

- The space in the title block is divided into- Identification zone and Additional information zone.
- The **identification zone** includes the following basic information:-
    1. Registration or identification number
    2. Title of the drawing
    3. The name of the legal owner of the drawing, i.e., name of the firm or the company
- The **additional information zone** includes the following items:-
    1. Indicative items
    2. Technical items
    3. Administrative items

The indicative items are: The symbol indicating the system of projection employed for the drawing, the main scale of the drawing, and the linear unit of the dimension if it is other than the millimeter.

The technical items are: Method of indicating the surface texture, method of indicating the geometric tolerances, values of general tolerances and the other relevant standards.

Method of Numbering the Multiple Sheet Drawings

- If more than one sheet is required for a component or an assembly, then the same registration or identification number must be indicated by means of a sequential sheet number.
- In addition, the total number of sheets should be shown on sheet 1, for example, Sheet No. n/p. Where, n is the sequential sheet number and p is the total number of sheets.

  Instruments required for measurements:
    1. Measuring tape
    2. Cross stop
    3. Ranging rod
    4. Pegs and bars

Landscape (garden) symbols:

- Garden symbols are nothing but the pictures, which are used for representing the various garden components (features). They are mainly used in preparing the garden plan or design.
- The symbols cover primarily the garden surfaces such as steps, different types of shrubs, trees, climbers, perennials and other plants.
- On plan, the symbols are drawn to scale wherever possible and make the transition from broad outline plan to detailed structural and planting plans.
- When drawing symbols, generally it should look as much as possible like the shape of the plant or structure it is representing.

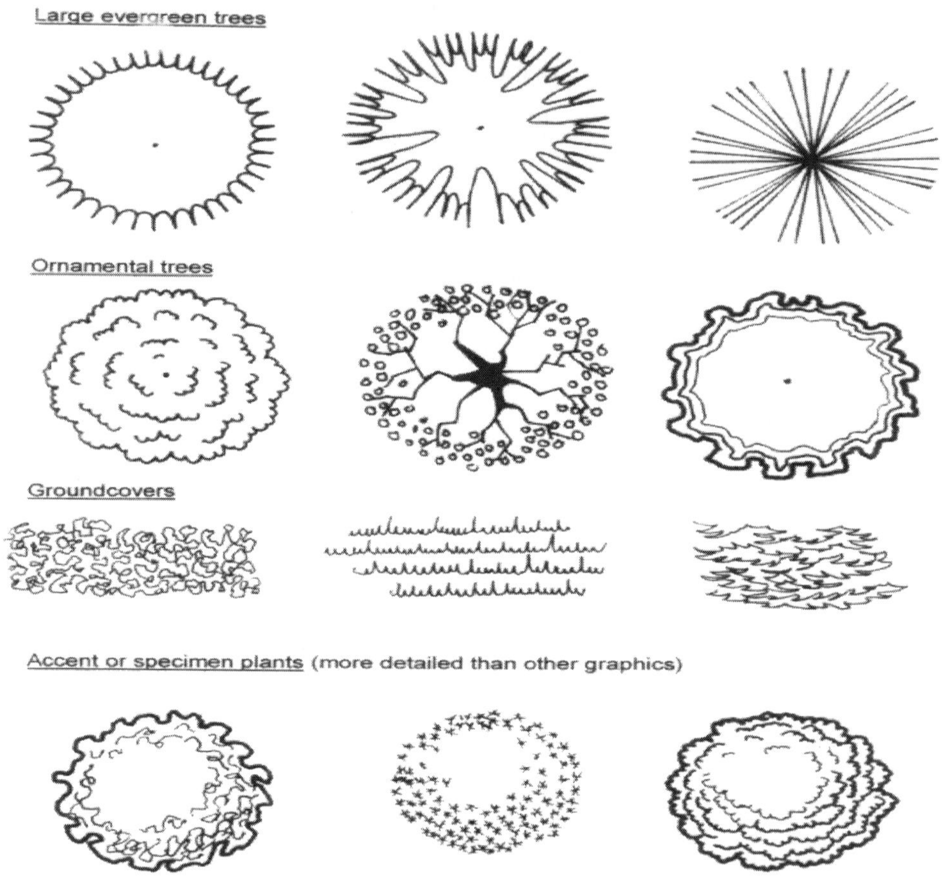

Reterences

1. Randhawa, G.S. and A. Mukhopadhyaya, 1986. Floriculture in India
2. Bhattacharjee, 2004. Landscape Gardening and Design with plants.

Chapter **5**

Tools and Implements of Landscape Designing

Although many tools and implements may be used for a single horticulture operation, use of a right type of tool for a particular operation to achieve maximum efficiency and quick turnout of work is indisputable. Therefore selection of suitable tools/implements is essential for carrying out various horticultural operations. For example several tools are there for cutting branches of plants such as secature, tree pruner, pruning saw, carpenter's saw etc. If the thickness of branch is about 1-2 cm, the secature may be more suitable, similarly if the branch in 4-6 cm is thickness pruning saw is the proper tool and if the branch is still bigger (more than 8cm) carpenter's saw may be required. Therefore one has to exercise option to select a suitable tool/implement to carryout various operations from the stage of land preparation to harvesting.

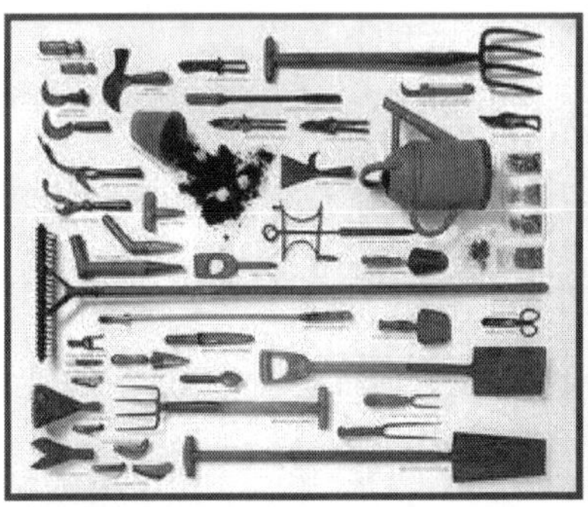

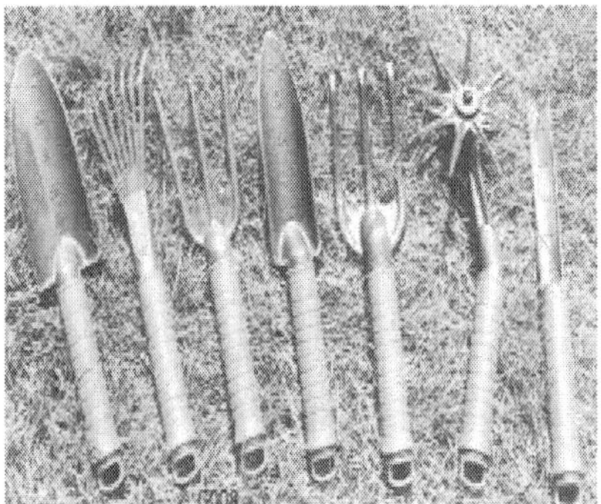

Tools, implements and plant protection equipments are commonly used in the horticultural crops are as follows:

Tools and Implements

Sl. No.	Name	Uses
1	**Axe**	For felling the trees
2	**Bill hook**	To cut the bigger stems near the ground surface
3	**Budding-cum-grafting knife**	For both grafting and budding. It has two sharp blades to give cuts on the stock and scion and the back end (flap) made up of plastic or brass used to lift or loosen the bark for inserting the bud.
4	**Pick Axe**	For light digging and loosening the soil
5	**Carpenter's saw**	To prune the bigger branches (more than 8cm diameter). It is especially useful in crown grafting
6	**Crow-bar**	An iron rod of 4-6 ft in length with one end pointed and the other end flattened. Used for digging pits and moving rocks
7	**Digging fork**	To loosen the moist soil or manure pits
8	**Dutch hoe**	To loosen the surface soil between the rows of plants for removing small weeds
9	**Drainage hoe**	To make the drainage channel and to remove silt deposit in the channels
10	**Dibbler**	To make small holes on the seed beds in order to place seeds or transplant seedlings
11	**Forester's shear**	To prune the medium sized branches (4-8cm) which are at higher height on the trees
12	**Garden hand rake**	For removing stubbles, small stones, leveling of nursery beds and formation of small beds
13	**Garden trowel/ shovel**	For lifting more number of seedlings
14	**Grass shear**	To cut the out growth of grasses planted in posts, carpet beds
15	**Hand fork**	To loosen the soil in seedbeds and to break the clods
16	**Hand kudaali**	For light digging and other intercultural operations
17	**Hand cultivator**	To loosen the soil, remove clods, pebbles in nursery beds and mixing of manures and fertilizers
18	**Hand weeder**	To remove weeds from soil
19	**Hatchet**	To remove or cut down the bigger stems and broken stems
20	**Hedge shear**	To prune the tender parts of garden shear the plants, it is especially useful for trimming hedges, borders, topiary work
21	**Hose pipe**	To irrigate flower beds, lawns etc
22	**Khurpi**	For weeding and stirring the soil in the pots and beds
23	**Lawn mower**	To cut the grass uniformly in the lawn. It is having a roller behind to pad the grass to have cushion

Sl. No.	Name	Uses
24	**Lawn sprinkler**	For irrigating lawns
25	**Pick-axe**	For opening trenches, channels, digging pits, loosening soil
26	**Pruning saw**	To prune the thicker branches (4-6cm girth) of an acute crotch (angle)
27	**Pruning knife**	For pruning of thicker branches and it has curved knife
28	**Pruning shear**	For cutting small sized branches
29	**Rotary weeder**	For cutting of grasses in lawn, carpet beds, edges
30	**Secateurs**	To prune the branches, twigs, water suckers etc. of small plants
31	**Sickle**	For cutting grasses, vegetables
32	**Scythe (Dabba)**	For cutting lawn grasses, vegetables
33	**Spade**	To loosen the soil, prepare irrigation channels, collect the soil in heaps and facilitate filling up of soil, manure in the baskets
34	**Transplanting trowel**	To lift the young seedlings along with a boll of earth for transplanting
35	**Tree pruner**	To cut down the smaller branches of the trees without climbing
36	**Trenching hoe**	For light collection of soil, irrigation purposes and opening of trenches
37	**Tree calipers**	To measure the girth of trees trunks
38	**Water can with Hose head**	To water the young seedling in seed beds, potted plants etc. the rose head facilitates with fine spray of water which prevents the washing down of soil
39	**Weed cutters**	Special kinds of weed cutter have a serrated double edged steel blade and handle about 60cm long. It is used with swinging strokes in two directions
40	**Wheel barrow**	To transport manures, soil, seedlings, garden waste
41	**Back pak sprayer**	For spraying nursery beds, flowerbeds, potted plants
42	**Power sprayer**	For spraying plant protection chemicals on large areas
43	**Knapsack sprayer**	For spraying nursery beds, flower beds, potted plants
44	**Automizer**	For spraying nursery beds, potted plants
45	**Hand Rotary Duster**	For dusting the powdery chemicals on plants
46	**Hand sprayer**	For spraying potted plants
47	**Rocker sprayer (Gattar)**	For spraying plant protection chemicals particularly in plantations/orchards
48	**Foot Sprayer**	For spraying it requires two labours for operation

References

1. Bose, T.K., D. Mukherjee, 1977. Gardening in India
2. Randhawa, G.S. and A. Mukhopadhyaya, 1986. Floriculture in India
3. Arora, J.S. 1998. Introductory ornamental Horticulture
4. Bhattacharjee, 2004. Landscape Gardening and Design with plants.
5. www.gardening.about.com

Chapter **6**

Styles of Gardening

Before someone starts to venture into designing a garden it will be wise to get an idea about the major gardening styles of the world. This will open up a window to this knowledge on gardening and help him design his own garden by adapting the best from each or any of these. But this does not mean that one should copy any garden style. A garden enthusiast has to study the different styles of gardening only to gain knowledge to help him from his own ideas suiting the local condition and limitations such as a available space, funds, etc. Though in India from history and ancient literature we find that gardening was quite in vogue in olden times, but unfortunately there is no garden style called "Indian garden", which can claim a place in the major gardening styles of the world. The famous garden style of India the "Mughal Gardens" are nothing but a replica of the ancient Persian Gardens.

There are THREE main styles of gardening i.e. Formal garden, Informal garden and Free or Wild garden

1. Formal Gardens

A formal style garden is laid out in a symmetrical or a geometrical pattern. For this, first the design is made on a paper and then land is selected accordingly. In this garden style, the design is stiff as everything is done in a straight and narrow way. In such gardens everything is planted in straight lines. Also if there is a plant on the left hand side of a straight road, a similar plant must be planted at the opposite place on the right hand side i.e., mirror image of each other. The flower beds, borders, and shrubbery are arranged in geometrically designed beds. Trimmed formal hedges, Cypress, Ashoka trees, and topiary are typical features of a formal garden. Eg. Mughal, Persian, Italian and French styles

2. Informal Garden

In an informal style garden, the whole design looks informal, as the plans and the features are arranged in a natural way without following any hard and fast rules. But here also the work has to proceed according to a set and well-through-out plan; otherwise the creation will not be artistic and attractive. The idea behind this design is to imitate nature. Eg. English and Japanese gardens

3. Wild Garden or Free Style Garden

It is a comparatively recent style of gardening, namely, "Wild Garden" was expounded by William Robinson in the last decade of the nineteenth century. His idea was revolutionary and found many admirers to follow this. The concept of wild garden is not only against all formalism but it also breaks the rule of landscape styles. His main idea was to naturalize plants in shrubberies. He also preached that grass should remain unmowed, as in nature, and few bulbous plants should be grown scattered in the grass to imitate wild scenery. He also suggested that passages should be opened in the woodland, and trees, shrubs, and bulbous plants should be planted among the forest flora to fulfill his idea of a wild garden. His other idea was to allow the creepers to grow over the trees naturally imitating those of the forests.

References

1. Bose, T.K., D. Mukherjee, 1977. Gardening in India
2. Randhawa, G.S. and A. Mukhopadhyaya, 1986. Floriculture in India
3. Bhattacharjee, 2004. Landscape Gardening and Design with plants
4. Randhawa, 1961. Beautiful trees and Gardens; IACR, New Delhi.
5. Arora, J.S. 1998.Introductory Ornamental Horticulture

Chapter 7

Formal Style Garden

The major Formal style gardens are:-

1. Italian gardens
2. French gardens
3. Persian gardens
4. Mughal gardens

1. Italian Gardens

- The Italian garden style came into existence at the Time of Renaissance.
- There is a striking similarity between the Persian and the Mughal styles with the Italian style.
- In all these styles of gardening the similarity was the use of heavy masonry features, through the character of masonry was different in the Italian style.
- The Italian elites conceived their gardens just as an extension of the lavish palaces, as a glamorous outdoor hall for entertainment and for showing off their wealth as well as status.
- The most prominent features associated with Italian gardens was the massive flight of stairs, generally of marble, complete with balustrade to connect the different levels in the garden.
- Decorative urns, fountains generally in combination with stone sculptures or rather the fountains themselves used to emerge from one part or the other of the statues, are the other equally important features of the Italian gardens.
- The greatest benefit the Italian garden style brought to gardening is that it taught all of Europe that gardening could be a most respectable form of art which demanded through and careful study.

2. French Gardens

- In the sixteenth-century France, the court life was shrouded with stiff formality and exactness. Matching with that the French style of garden designs were also very intricate and artificial.
- Until this time the French gardens were nothing but copies of Italian style.
- The new style of gardening now known as French style is largely, rather solely, due to the efforts of Le Notre who served in the Royal Garden of Louis XIV from 1643 to 1700.
- He elevated the art of garden design to a standard which had never since been reached. It was Le Notre who showed to the world the impact in impressiveness of scale, on garden design.
- His main creations, the gardens at Versailles, have avenues which are memorable for their tremendous length and width.
- To design a garden at Vauxe-le-Vicomte, his first master piece, Le Notre had removed three villages to create his vista. The moral of French garden style of Le Notre seems to teach the lesson "How to Think Big".
- The style of Le Notre can be termed as an evolution and mastery of the art of formal garden in its perfection. His style dominated the gardens of civilized Europe, for a long time.

3. Persian Gardens

- Persian garden style is one of the oldest. The Persian garden style and the Japanese style both were based on their respective ideas of heaven. Except this similarity the two styles differ widely from one another.
- The Persian styles were strictly formal and symmetrical. They used for their gardens all crafted materials such as masonry, carved and pierced marble stones, and highly polished stones.
- The Persian gardens were laid out by cutting terraces on the hill slopes. They also tapped some natural spring to create a straight water-course through the gardens and manipulated the water-course to undertake different movements along its run.
- If there was no natural source of water this was created artificially by diverting some rivulet or a stream. So the main stay of a Persian garden design was nahars (flowing canals) of water – the concept of Persian Paradise, "where cooling water flows".
- The selection and placing of trees were very judicious. The fruit trees represented the symbol of life while Cypress symbolized death and eternity.

4. Mughal Gardens

- The gardens laid out during the rule of Mughal Emperors in India are known as Mughal gardens. The Mughal gardens are similar to the style of gardens of Central Asia and Persia.
- Babar (1494 – 1531) was the first Mughal ruler to introduce this style in India. All other Mughal rulers and some of the Mughal Begums starting from Akbar and followed by Jahangir, Nur Jehan, Shah Jehan, and Aurangzeb all laid their hand on developing one or the other Mughal gardens in India.
- The main features of Mughal gardens, which are largely borrowed from the Persian style.

Main Features of Mughal Gardens

1. Site and Design

- Mughals were very choosy about the selection of site and always preferred a site on a hill slope with a perennial rivulet or along the bank of a river.
- Mughal gardens are generally rectangular or square in shape and different architectural features are the main stay of the design.

2. Walls and Gates

- The Mughals created the gardens not only for pleasure and recreation but also as forts and residences surrounded by high walls and with-an-imposing wooden gate at the entrance which was studded with bold iron nails and pointed iron spikes.
- The purpose of the high walls was security from the enemies and shelter against hot winds. The gardens were a place of peace for the Emperors to enjoy with their wives and concubines.

3. Terrace

- The Mughals came from the hilly terrains and so they were fond of terraces in the gardens.
- For this reason they used to select the location of gardens near hill slopes.
- Their fascination for terrace was so intense that even in the plains of Punjab they created artificial terraces.
- According to Islamic faith the Paradise has eight divisions and hence sometimes the gardens have eight terraces

4. Nahars (Running Water)

- The style of having running water by constructing canals and tanks borrowed from the Persians.
- The water canals were paved with tiles (or marble) of blue colour to create the illusion of depth.
- The course of water used to be maneuvered in various ways taking advantage of each slope, however small it may be, to break up the flow into artificial falls and ripples.
- The water canals used to have fountains to throw up the water high in the air. In the evenings small lamps used to be illuminated to create beautiful reflexions.
- The fascination for water came from the Muslim faith which says that the promised paradise is the place "where cooling waters flow". With this bias for water Mughals selected sites on hill slopes to tap a running rivulet or spring for source of water.
- In the plains of India, where the summers are hot, Mughals utilized water for its cooling effect.

5. Baradari

- This is noting but an arbour-like structure, but made of stone and masonry with a pucca roof and a raised platform for sitting.
- These were usually provided with twelve or occasionally more doors on all sides for the Emperors to watch the performance of the dancing girls.

6. Tomb or Mosque

- It was a common practice to have the gardens built around a tomb (e.g., Taj Mahal, Akbhar's Tomb at Sikandra).
- It is often said that the Mughal gardens were at their best when built around a monument.

7. Trees and Flowers

- The trees were selected with careful planning and thought, as to Mughals each tree symbolized something, like life, youth, death, etc.
- Fruit trees were considered symbols of life and youth, while, Cypress represented death and eternity. The Mughals had bias for spring flowering trees and flowers.
- The seasonal flower beds were of geometrical pattern and constructed along the water canals or near the main buildings.
- The favorite flowers were rose, jasmine, carnation, hollyhock, delphinium etc.

References

1. Bose, T.K., D. Mukherjee, 1977. Gardening in India
2. Randhawa, G.S. and A. Mukhopadhyaya, 1986. Floriculture in India
3. Bhattacharjee, 2004. Landscape Gardening and Design with plants
4. Randhawa, 1961. Beautiful trees and Gardens; IACR, New Delhi.
5. Arora, J.S. 1998. Introductory Ornamental Horticulture

Chapter 8

Informal Style Garden

The major Formal style gardens are:-

1. English gardens
2. Japanese gardens

1. English Gardens

- Due to favorable climatic conditions such as high annual rainfall, the natural ground cover in the English countryside is grass.
- With this in mind the famous British garden architects Repton and Capability Brown advocated the concept that the British gardens should like the countryside. Their main idea was that the gardens should merge with the countryside without any artificial barriers such as fences, hedges, or the like.
- But it is only in the 18th century that these two gentlemen along with Kent brought the touch of nature in the garden, although the history of gardening in England dates back of fourteenth century.
- They started kitchen gardens to supply vegetables to the inmates of the monasteries and grew herb gardens for medicines. But this gradually imbibed the spirit of gardening to the people who realized the goodness of residing in pleasant surroundings.
- By the middle of the sixteenth century the English gardens saw flower beds, topiary, and terrace gardens.
- In the middle of the eighteenth century gardens were laid out with more emphasis on architectural features.
- The main features of gardens during this period were curved paths, informal groups of trees, rivulets or streams, artificial waterfalls, and clipped hedges.
- The flowering annuals, the main stay of an English garden, came into prominence during the nineteenth century which subdued the architectural features.
- *The main features of an English garden are* the lawn, mixed border especially of herbaceous annuals, as well as herbaceous perennials, shrubbery and rock gardens. The English climate suit admirably well for the growth of herbaceous annuals.

- This prompted them to evolve a large number of hybrids of annuals as well as to collect the best flowering plants from all over the world.
- Most of the flowering annuals that we see today in the Indian gardens, with few exceptions of Amaranths, Balsam, Gomphrena, Marigold etc., were brought here by the British.

2. Japanese Garden

- The Persian and Japanese garden designs were based on their respective ideas of heaven. One most admirable feature of the Japanese garden is that while other major styles of gardening of the world changed radically or fallen into dis-flavor, the Japanese continued the same style for centuries but still remained popular. This can be attributed to the special relation of the Japanese gardens to nature.
- A most important teaching of the Japanese garden is possibly that "unless a garden has an air of peace it's not worth a place visiting. It should be a place where the mind finds rest and relaxation." The Japanese gardens which are planned with so much of thought appear so casual that they do not appeal to many people.
- Another strong basis of the Japanese garden is its immutability, i.e., except some seasonal changes in the deciduous trees, the Japanese garden hardly goes through any other strong visible changes during different seasons.
- The immutability is achieved also because rather than a grand mixture of flowers, shrubs and trees more emphasis is placed on natural elements such as a simple path, a group of rocks, stepping stones, streams, waterfalls, bridges, stone lanterns, and so on.
- A Japanese garden tries to capture natural scenery or to imitate a landscape. The three elements most important to achieve these objectives are water, stone, and plants.
- Low sculptured bushes and trained dwarfed trees look very attractive in a Japanese garden.

Types of Japanese Gardens

The Japanese gardens are further classified based on positions, shape, and purpose. The important types are as follows:

a) **Hill Garden**

- This style is known in Japanese as 'Tsukiyama-niwa, meaning hills and water. The hill garden is made up of one or more hills designed with earth mounds and exposed weathered stones.
- The other features of this garden area water in the form of a stream or a pond or waterfalls or all the three with or without islands and also bridges, lantern, stones, and trees.
- The important points in the garden are decorated with stones and selected trees. But pine trees may be planted to give the effect of being swept by wind.
- Untrimmed stepping stones are placed over the walks. An island is generally a usual feature in a hill garden.

b) **Flat Garden**

- As the name implies, Hira-niwa or flat gardens are laid out in flat ground without hills or ponds. Flat gardens are supposed to represent a mountain valley or a meadowland.

- These gardens were popular during the era of Muromachi (1392-15723).
- A Fat garden is not necessarily as flat as a pan-cake. Since it stimulates a mountain valley, low rounded hills designed with the help of stones or earth mounds or both will look quite appropriate in a flat garden.
- The usual features to break the monotony of a flat garden are a well, a water-basin made of stone in the shape of an urn, stones lying close to the ground. In a flat garden, the principle is to avoid strong vertical lines represented by tall pines.

c) **Tea Garden**

- The tea garden is laid out based on certain principles and customs of the Japanese tea ceremony and hence needs a considerable space of at least about 200 square meters, for its designing.
- Since the performance of the tea ceremony needs an atmosphere of intimacy, it is essential that the garden be enclosed by a fence.
- But the fence should be rustic in mature, with a gate made of very light material such as bamboo.
- To protect the tea house from the noise of the outer world, the tea gardens are divided into an **outer garden (soto-roji)** and **inner garden (uchi-roji).**
- The outer tea garden is comparatively a narrow area, with a waiting place where the guests are supposed to wait until the master of the house appears to welcome them.
- The inner garden contains the tea house. The tea house of the classical time was nothing but a small straw hut with an outside waiting place, a small side room for washing the utensils, and the main ceremonial tea house itself having a capacity to accommodate only five persons.
- The most important feature at the entrance of the tea house is a water-basin or a well or both for the visitors to rinse their face before entering for the tea ceremony. To illuminate the water basin and resting place stone lanterns are set in appropriate places.

- The selection of water-basin and stone lanterns is done with scrupulous care so that both the purposes of utility and beauty and elegance are fulfilled.
- As far as tree planting is concerned, a contrast is maintained between the outer garden and the inner garden. The outer garden will have simple plantings and stone groupings.

The entrance to the tea house is through a low-door so that the guests have to enter in a bending posture, simulating respect and humility.

d) **Passage Garden**

- The passage gardens, the Roji-niwa, are those which are laid in narrow passage, as for example a narrow space between two houses or approaches to buildings.
- As such areas are generally narrow, the garden lay-out should be simple and not over crowded.
- In such gardens there should be hardly any ornaments such as lanterns, basins or other man-made features. The common features of a passage garden are a few key rocks, slabs of stones, and only a couple of types of plant.
- Bushy shrubs and trees are unsuitable in a passage garden; instead, plants with open form and slender shapes are selected.
- The passage gardens generally occur as an accidental necessity in a narrow passage formed by two buildings of a large property. But, occasionally a narrow passage is purposively created to simulate a distant prospect.

e) **Sand Garden**

- It is the simplest style of gardening, though not liked by many as it is totally devoid of plants. The main feature of this style of gardening is to arrange few vertical and prostrate stones in groups of 2 or 3 and to fill in gap between the stones with fine white gravel.
- The gravel is raked in most simple patterns simulating the ripples of flowing water.
- The raking has to be repeated often to keep the garden in its best shape. This style of garden looks pleasant and effective only when confined to a limited area.

English Garden

Japnese Garden

Japnese Garden

Sand Garden

Important Features of Japanese Garden

1. **Ponds**

- The geometric shape does not live up to the Japanese tradition and hence all ponds are of irregular shape.
- The banks are generally bordered with stone piling work in a regular or an irregular fashion. In olden days the bottom used to be made impervious by puddling clay.
- But now-a-days concrete-bottomed ponds are also used. The ponds are generally fed by a stream or a waterfall.

2. **Streams**
   - Small streams are arranged most naturally with natural stones bordering the banks. The flow of water in the stream may be manipulated by changing the inclination.
   - Sometimes it may be as fast as a mountain rivulet or like a smooth-flowing river of the plains.

3. **Waterfalls**
   - The presence of waterfalls in Japanese gardens is another attempt in imitating nature. A waterfall may be made more effective by manipulating it to drop in two or three levels.
   - To make the waterfalls dignified and natural, large-sized stones are arranged around these. The dense planting of evergreens around the waterfall also symbolizes mountain scenery where actually a waterfall naturally belongs.

4. **Fountains**
   - Often natural fountains are provided near the foot of the hill, on the hillside or in the forest. Often water is conveyed from a hillside by means of bamboo-piping.

5. **Wells**
   - In olden days wells in the garden were features of utility. In modern times this feature is more in the nature of an ornament than a utility.
   - The wells may be square, circular or criss-cross in shape. The frame is generally constructed of stone but occasionally wood is also used.
   - The frame is generally raised above the ground level to a height of 45-60 cm.

- The mouth of the well is often covered with a mat or a flexible lid knitted out of bamboo. Square-shaped buckets are hung on either side of the pulley, suspended from a rope.
- The well must be made very ornamental.

6. **Islands**

There are four important garden islands in Japanese style, the first two types representing Sea Islands.

1. ***"Elysian Isle"* (Horai-jima)"-** It is constructed in the middle of a lake and is never connected by a bridge. The beach is decorated with sand from the sea, shells, and pebbles. No fresh water vegetation is planted over this island. Often this island is given the shape of a tortoise.
2. ***"Wind'swept Isle"* (Fukiagae-jima)-** It is also an ocean isle and constructed in a similar way. In addition to sand and shell, the beaches here are decorated with sea rocks.
3. ***"Master's Isle"* (Shujin-to)-** It is placed in the foreground of the landscape so that it can be easily approached by a bridge from the bank. The "Master's Isle" generally has a summer house which is nothing but a thatched arbour. Besides the arbour, some selected trees, stones, stepping-stones, and one or two lanterns are included. Often the islands are in the shape of mounds or hillocks. Often the islands are made a peninsula, connected to the mainland by a narrow neck of land instead of a bridge.
4. ***"Guest's Isle"* (Kiakujin-to)-** It is located in the background and is accessible by bridges and stepping-stones.

7. **Bridges**
   - These are used to reach an island or for crossing a stream. Bridges may be made of stones, polished or un-worked earth, wood, and other materials.
   - The Japanese concept of a bridge is not just a quick crossing of a water way. The aim is to prolong the crossing time so that the visitor gets enough time to enjoy the scenery around.
   - The earthen bridge is constructed by first putting a wooden framework over which bundles or small logs are laid across and then these are covered with about 15 to 20 cm layer of earth and gravel.

8. **Water Basins**
   - The water-basins are fitted near a house meant for the guests to rinse their mouth and wash the hands. But in present days these basins have become nothing but ornamental features.
   - A small house may have a basin one meter tall; whereas in front of a large house the basin becomes as tall as 2 to 2.5 m and thus making useless as a place of washing, but remains there as an ornamental feature. The basins are generally fitted with an ornamental lid.
   - The water-basin comes in various shapes, the most common ones are in the shape of an urn. But square-star shaped, cylindrical, stone-bottle shaped, and bowl shaped basins are also quite common. A screen-fence is provided near the water-basin to screen off unwelcome sight.

- Stones are placed at the base of the water-basin to arrest the splash of water which otherwise may wet the space below the veranda. A lantern is provided nearby for illumination.

9. **Stone Lanterns**

- Stone lantern is an important feature of any Japanese garden. The usual stone used is granite, but sand stone or white marble may also be used. The latter one is specially used for "Snow-Scene" type of lantern.
- The usual places of fixing the lanterns are near the base of a hill, on an island, on the banks of a lake, near a water-basin or a well, along a path, on a boat-landing, near waterfalls, and a bridge.
- The lanterns are used singly but along with a combination of rocks, fences, water-basins, shrubs and trees.
- The lanterns are not exactly meant for illumination but as objects of ornaments. Even when a lantern is lit in the night, the light emitted hardly illuminates the place because mostly a group of bushes are grown near it. The object of illumination then becomes a dim mysterious glow.
- When a lantern is near a river bank, this is lit to have red reflection on the water surface.
- A lantern has six parts, namely, the ornamental top, cap, light chamber, middle stand, post, and base. These may be of various sizes and shapes. The size should be in proportion to the building or the other garden features around the lantern.
- The first lantern was erected in Japan in the seventh century by the son of Emperor Suiko. Lanterns looking of age are valued much and may people make the lanterns look old by artificial means. One such method is to attach moss to the lantern.

10. **Stones**
    - Stones are selected according to size, shape, and colour.
    - No stone should be disproportionate to the size of the garden or the features around it, like a lantern or a water basin.
    - The stones are rarely placed in isolation but rather arranged in groups of two to five. Each group will have stones of various heights and shapes.
    - All stones must be arranged with a firm foundation as stones of unstable nature show the weakness of a garden design.
    - The principle governing the arrangement of stones is to make them look natural. For this purpose, low-growing bushes or upright trees are planted near the stone groupings depending upon the size of the nearby stones.
    - Naturally, the size, shape, and colour of the stones vary according to the purpose and the place of their use. Accordingly, the stones on a hill, on the blanks of lakes or streams, and near water cascades vary in their shape, size, and ornamental features.

11. **Pagodas**
    - Another favorite feature of Japanese landscape is the stone tower or the pagoda which is a structure consisting of two, three, five, or more separately roofed stages.

12. **Fences and Gates**
    - Fences in a Japanese garden are of two types, one is for partition, while the other is for the purpose of screening which are called "*Sleeve Fences*". The latter type is so named as its shape resembles that of the long sleeve of a lady's Kimono.
    - The partition fences should look light in appearance and hence wood and twigs of bamboo are preferred to stone.
    - The screen fences are meant for covering something which is not pleasant in sight. This is made of wood or bamboo (whole or split or twigs) woven into patterns thus giving it an artistic look so that it can stand on its own in the garden as a feature of ornamentation.
    - There are generally two gates, one is the front entrance and second the back entrance. Gates are also made of light materials such as wood or bamboo. Some gates are bare while others are roofed. The roof may be made of bamboo, wood, or simply thatched.

13. **Vegetation**
    - The initial step in a Japanese garden is to decide the contours of land and water.

- Then comes the arrangement of principal rocks which are of primary importance in a Japanese garden as these forms the basic structure or the skeleton of the garden.
- After the stones, the secondary garden framework, i. e, the evergreen plants, are arranged. The trees are of permanent nature and stand as reference points in the garden.
- Time and again it has been told that the aim of a Japanese garden is to imitate nature by using natural elements, and hence, there is hardly any bar in using any plant material which serves this purpose. The Japanese use both needle-leaved and abroad-leaved trees in their garden.

Some Typical Trees of Japanese Gardens are:

a) **Evergreens**: Pines, different species of *Abies, Cryptomeria japonica, Podocarpus macrophylla*, and *Juniperus chinensis*.

b) **Deciduous**: Maples (Acer species), Poplars (*Populus* sp.) Mulbery, (*Morus alba*), and *Salix babylonica* (willow)

c) **Flowering:** The most commonly used plants are different *Prunus* species, besides *Magnolia grandiflora* and others.

- **The commonly grown shrubs are**: *Aucuba japonica, Azaleas, Gardenia florida, Nandina domestica, Camellia, Lagerstroemia indica, and Rhododendrons.*
- **Bamboos** play a special role in the Japanese gardens. The striking patterns of shadow cast by the arching bamboos against paved path, fences, and patios look beautiful. A paved path in the entrance garden bordered by bamboos simulates a grove.
- **The flowers**: The Japanese use more commonly are: Chrysanthemums, Asters (e.g., *Aster fastigiatus, A. glehnii, A. microcephalus*), Carnation, different Lilies, Irises, Lotuses, Peonies, and Orchids.

- **Among the vines,** (*Clematis, Lonicera japonica, Ipomoea hederacea (Syn. Pharbitis headeacea), I. Purpurea (Syn. Pharbitis purpurea) Trachelosermum jasminoides,* and *Wisteria sinensis* are often used.

References

1. Bose, T.K., D. Mukherjee, 1977. Gardening in India
2. Randhawa, G.S. and A. Mukhopadhyaya, 1986. Floriculture in India
3. Bhattacharjee, 2004. Landscape Gardening and Design with plants
4. Randhawa, 1961. Beautiful trees and Gardens; IACR, New Delhi.
5. Arora, J.S. 1998.Introductory Ornamental Horticulture

Chapter **9**

Special Gardens of Landscape Designing

Shade Garden

- Making of garden under the shade or in shady area is referred as shade garden. During gardening under the shade of trees, the tree roots compete with the plants grown there for food and water. So, the soil under the trees is to be dug deeply to about 25-30cm to cut the surface feeding roots. This operation should be repeated every year by digging deep along the borders of the beds. The plants are to be watered and fertilized separately.
- The shade garden plants under trees are spaced much wider than the open. A shaded place in between two buildings can be developed into a wild garden with a winding path.

Plants suitable for shade garden are- Azalea, Hydrangea, Hill balsom, Coleus, Trade centre, Pothos, Syngonium, Ipomea, Vadlia, Peperomea, Bulbous ornamentals etc.

Sunken Garden

- When a garden is laid below the ground level it is known as sunken garden. An existing ditch or a dried out pond may be utilized for laying out a sunken garden or a portion of a garden may be dug to layout such a garden.
- A provision should be made for proper drainage. The surface runoff water should not fall into the sunken garden. A formal or an informal garden can be laid out in a large sunken garden.
- The ideal depth of a sunken garden is about 120cm. The approach to a sunken garden is generally by a staircase preferably made of stones. The

surface of the sunken garden can be decorated with crazy paving, gravel path etc.

Plants suitable for sunken garden are- Ferns, Ireora, Gardenia, Flowering annuals, lawn grass, Russelia etc.

Terrace Garden

- A terrace is a raised space of ground constructed around a dwelling house or at the corner of a garden or on the sides of a hill. When this terrace is used for some sort of gardening this is known as Terrace Gardening.
- Creating garden on the roof of the building is known as roof garden. In modern days homes with compound and lawn especially in cities and towns are becoming rare.
- As a result the private home gardens are vanishing and well-planned roof gardens can be a place of joy and recreation. This is basically meant for a place of leisure and pleasure.
- A terrace garden is generally constructed just in front of the house from where a view of the whole garden can be obtained. But this can also be constructed in any other corner of the garden from where a clear view can be obtained.
- The fundamental difference between ground gardening and roof gardening is that at found level the depth may be limited with source of ground water. Whereas on the roof, the depth of the soil is shallow not exceeding 60cm but generally varying between is 20 to 30 cm.
- Container plants may be kept on verges or bricks. Over the roof adequate drainage should be provided for the rain and excess irrigation water to drain off quickly. Water garden, rock garden, lawn, hanging baskets, climbers may form a part of roof garden. Always allow for the vertical growth rather than the horizontal growth.

Plants suitable for Terrace Gardening are: Only a few low-growing creeping plants such as in Tropics *Portulaca, Lantana sellowiana* and in temperates Thymus and *Veronica repens*. Some small specimen dwarf trees or shrubs can be planted.

Rock Garden

- Rocks and soils are arranged in such a fashion that conditions are created for the growth of different plants. Such type of landscape designing is known as rock garden.
- In nature, rocks may be getting covered with different colored moss and if soil collected in cracks and crevices of rocks, plants grow there naturally.

Selection of Site

- It is easy to create rock garden where natural rocks are available nearby. But large rock garden should be situated in a place where there will fit with the other garden features.
- Apart from this corners, middle part of informal garden, under large trees are better places for creating rock garden.

Characteristics of Rocks

- The rocks should be of local origin, porous and have a weathered look.
- Stones having diameter of 60cm or more should be selected. As well as the rock plants growing in rocks itself.

  Some Common Types of rockery:

1. Tyre type of rockery
2. Round rockery
3. Rectangular rockery
4. Square rockery
5. Rockery under tree
6. Alpine rockery
7. Flat type rockery

Construction Procedure for Rockery or Rock Garden

1. First plan of the rock garden should be prepared on paper.
2. Selected site should be marked out by using bars and pegs.
3. The base should be filled with sufficient quantity of broken bricks, small stones and sand for proper drainage.
4. Form the mound of required height by filling sufficient quantity of soil, the mound should be allowed to settle for some time and formed by beating.
5. Create the steps, place the stones on the mound, the stone should be placed in a slightly slanting position gently leaning backwards, so that rain water and during watering, it flows towards the plant roots.
6. Sufficient number of pockets is created between stones for growing plants. While placing rocks side by side, they should be placed as close as possible otherwise soil will be washed away. Gaps between two stones should be covered with concrete. Concrete should be restricted to minimum.
7. Add garden soils up to a depth of 30cm. Mix together equal parts of sand and FYM.
8. Position the plants along with the pots so that one can see how they look and can move them around easily if necessary.
9. Use a trowel to take out a hole a little larger than root ball, plant is at correct depth, then tight loose soil around the roots and firms it well.
10. Finish off rockery by covering the surface by small pebbles, which increases the beauty and prevents soil erosion.

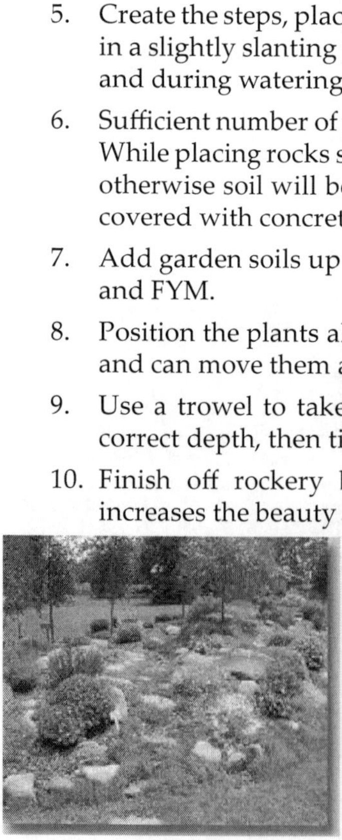
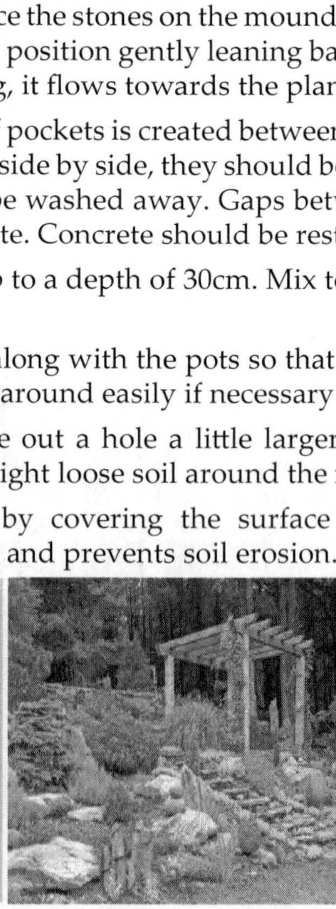
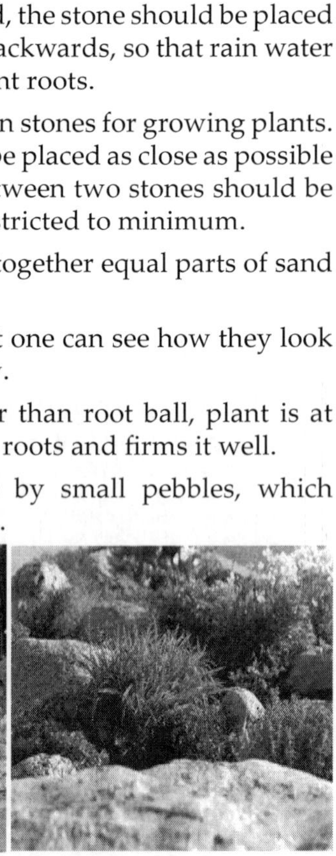

Different Features in Rock Garden

Maintenance of Rock Garden

- **Weeding:** Periodical weeding is necessary to remove the weeds, which are competing with plants. In addition to this all dried and dead leaves should be removed from the branches of the plants.
- **Stalking:** Stalking is necessary for newly planted tall plants to prevent the physical damages and to obtain the straight growth.
- **Watering:** Regular watering is necessary to keep the plants healthy and attractive. Watering should be done at least once in three days. Drip system of irrigation may also be adopted.
- **Replacement of soil:** Rockery plants need replacing of soil at least every 2-3 years as the soil become sour and infertile.
- **Manuring:** Plants may be top dressed every year with compost consisting of a good garden soil, sand and fine leaf mould in equal proportion to the top few centimeters of soil is scraped out and in its place new compost is filled in along with a little quantity (125g) bone meal depending upon the plants. Periodically rock plants should be fed with liquid manure also.

Suitable Plants for Rock Garden

- **Annuals:** All annuals may be accommodated in the rockery depending on their height in between shrubs and other perennial plants. Asclepias, Calliandra, Clerodendron, Crossandra, Cuphea, Euphorbia, Zinnia, Phlox, Verbena.
- **Shrubs:** Duranta, Juniferous, *Lantana sellowiana, Russelia juncea, Thuja orientalis* etc.
- **Cacti and Succulents:** Adiantum, Pony tail plant, Opuntia, Cerus, Mumalaria, Agave, Kalanchoe, Sanseveria, Yacca, Hawarthia, Furcarea, Sedum etc.
- **Ferns:** Nephrolepsis, Polypodium and Adiantum etc.
- **Shade plants:** *Imparatiens sultaniana, Pedilanthes tithymaloides, Pilea muscosa, Portulacca* sps, minature roses, *Zebrana pendula, Tradescantia albifolia, Vinca rosea* etc.
- In addition to above plants, with careful selection of large group of annuals, bulbous plants, other shrubs and other ornamental plants can be selected and grown in the rockery depending upon the necessity and environmental condition.

Gravel Garden

A garden which is made by using different types of gravel is known as Gravel garden. Gravel is an inexpensive and flexible alternative to paving of a lawn. It can be used in both formal and informal designs.

Types of Gravels:-Gravels comes in different sizes and colors. Some types are angular, others are rounded, some are white, and others are assorted shades of green or red.

Types of gravel garden: 1. Gravel Path 2. Gravel Bed

1. **Gravel Path**: It is often used as drives, but it is also good choice for informal paths with in the garden. However it is not a good choice for paths where you will have to wheel the mower.

   Procedure to establish Gravel path:

   1. Excavate the area to a depth of 15cm and ram the base firm
   2. Provide brick or stout edge to retain the gravel
   3. First place a layer of compacted hardcore. Add a mixture of sand and course gravel.
   4. Top up to the required height with the final grade gravel. Rake and roll the gravel repeatedly until the surface is firm and stable.

2. **Gravel beds:** Gravel can be used as straight substitute for grass and require much less maintenance.

   Procedure to establish Gravel bed:

   1. Excavate the area to required depth of about 15cm

2. Level the ground lay heavy duty black polythene or a mulching sheet over the area
3. Then spread the gravel on top and rake level
4. To plant through the gravel, draw it back from the planting area and make a slit in the polythene plant normally
5. Firm in and pull back the polythene before recovering with gravel.

Chapter **10**

Landscape Designing of Roof Garden

- In many articles the gardening on the roof is often termed as terrace gardening which is not strictly correct according to British practice. Terrace gardening is completely different from roof gardening.
- In modern times, homes with a compound and lawn especially in cities and towns are becoming rare replacing such homes.
- As a result, the private home gardens are vanishing and the only places left for gardening are the roofs of houses and the balcony. A spacious and well-planned roof garden can be a place of joy and recreation.
- In bigger cities of India, many of the large hotels and public buildings are developing this type of gardening. Even in many small cities now-a-days, a hotel has a roof garden with lawn, though not well planned. Many private houses in all over the country have roof gardens, including a lawn.
- The roof garden of Dr. B. P. Pal in Delhi, dominated by a collection of roses, besides other plants such as bougainvillea, pansy, etc., is a charm by its own right. There are many other private homes in our cities having charming roof gardens with rare collections of cacti, orchids, dahlia, chrysanthemum, and other plants.

Limitations with Roof Gardening

The difficulties encountered in a roof garden are completely different from that of gardens at the ground level. Important limitations are as follows-

1. Since the garden is at a high level from the ground, the cost maintenance may go up as everything has to be carried up.
2. But the more fundamental difference is in the depth of the soil. At ground level the depth may be unlimited with a source of groundwater, whereas on the roof the depth of the soil is shallow, not exceeding 90 cm, but generally varying between 20 and 60 cm.
3. Not only the depth is shallow in roof garden, but the drainage is also good and hence water has to be replenished constantly by frequent watering.
4. Large trees and shrubs are generally not grown on the roof as the growth of the tap root is limited by the roof below.
5. Moreover, because of shallow depth, good drainage and frequent watering, the plant food in the soil leaches off more rapidly and is to be made good by more frequent replacement.

Planning of roof garden- Before planning a roof garden certain points are to be checked.

- Flower beds can be made directly on the roof. This can be achieved by placing the soil between the outer wall or the parapet and a wall built on the inner side with the help of moisture-proof wood shuttering stones, or bricks. The width of such boxes will vary according to the available space, the nature of the plants to be grown, and other factors. Such beds directly constructed on the roof surface are not preferred always, as they are of permanent nature and may also damage the roof. Alternatively, the best method will be to arrange a series of containers holding soil, such as boxes, pots troughs, tubs, etc., These are placed over wedges or bricks to leave a clear gap between them and the roof for proper drainage and passage of air.
- The first thing to check before starting a roof garden is whether the roof surface can bear the weight of the soil. Soil, especially wet soil, is much heavier than one can usually conceive. Secondly, the roof should be made water-proof to prevent seepage of water into the room below and finally, it should have adequate drainage so that rain and irrigation water drains off quickly.
- The concept of roof gardening itself is artificial; therefore, while planning and designing this point should be kept in mind. As for example, it is not possible to create natural scenery with hills and valleys or natural forest scenery. As for any other gardening a roof garden should also have a dominant focal point, like lawn or water garden or rock garden or a prominent climber (e.g., Bougainvillea, *Clerodendron splendens, Bignonia venusta*, climbing rose).
- On a roof the space available for gardening is limited; therefore, it will be a good idea to encourage vertical growth. This is the reason why a roof garden lover should use more climbers and trailers in his garden, some of which can climb over a supporting pillar while others can grow rampantly over the boundary wall and spill over the top and hang down the side of the building. *Ficus repens* is a good root climber which can cover a wall. Other attractive creepers include *Cobaea scandens* (annual), railway creeper, *Vernonia elaegnifolia* (Curtain creeper), passion flower, *Thunbergia alata* (annual), etc.
- Hanging baskets should play a major role in a roof garden. Various types of basket either containing flowering or foliage plants can be displayed with great advantage at suitable places. One can also display with advantage the collection of bonsai and few alpine plants. One or a few vertical gardens can also be displayed in a planned manner in the roof garden.
- It is possible to grow a wide range of plant materials in a roof garden excluding, of course, the large shrubs and big trees.

Plants suitable for roof garden

Flowering Annuals: Antirrhinum, stocks, dwarf sweet peas, pansy, dahlia, chrysanthemum, marigold, sweet alyssum, phlox, calendula and verbena etc.

Herbaceous Perennials: Daisy, Canna, *Mirabilis jalapa*, Portulaca, *Solidago canadensis*, *Vinca rosea* and perennial verbena.

Shrubs: Many of the dwarf and medium shrubs can be grown.

Trees: One or two dwarf trees such as *Plumeria* sp., *Callistemon lanceolatus* and *Gliricidia maculata* can be grown as specimen plants. Some large to medium trees such as *Araucaria cookii, Brassia actinophylla*, etc., can also be grown till they are young.

Creepers: Different creepers that can be grown are

Bulbs: A variety of bulbous plants of annual or perennial nature can be grown.

Water plants: Water-lilies and other water plants can be grown in the lily pool in cement tubs.

Chapter 11

Landscape Designing of Paved Garden

- The foundation of the paved garden should be prepared in the same manner as for an ordinary pathway. Proper drainage should be ensured for the paved garden plants.
- A paved garden is one of the attractive features of a garden. There are some specific plants, which adapt themselves well to a paved garden.
- These should be dwarf in nature and stand a considerable amount of wear and tear from shoes of different weights.
- But a paved garden should be laid in a path, which is not used very often. A special paved garden may also be created if a path suitable for this is not available.
- Ordinarily a paved garden is meant for walking, although not very frequently and hence the interstices of the paved garden should be planted sparingly.
- After the foundation is laid, a good soil, preferably of lighter texture, is laid to depth of about 10 cm. Before laying the stones over this, 2-3 weeks' time should be given to the soils to settle.
- The stones selected should not be less than 25 cm in diameter; otherwise the paved garden looks clumsy. For paths, rectangular stones are more suitable, whereas for circles and squares irregular stones of any shape or size can be used.
- Between two stones gaps of about 4-8 cm are left and the stones are laid in an informal pattern. These spaces are filled with good soil and traces of bone meal are also added.
- The bounder of the paved garden is usually supported by any adjoining feature, these should be cemented as in the case of a crazy path.
- When the sides are supported by stones, the plants grown to cover these are allowed to grow a little inside the paved garden in an informal manner to give a more natural effect.
- Planting should be done with discretion and over-planting should be avoided. The crevices should not be filled completely, rather the plants should be placed irregularly either in small group or in isolated patches.

- A dense planting will look unnatural and will have very little appeal. In a sparsely planted paved garden even the worst clumsy visitor will be able to avoid tramping of a plant in flower.

Plants suitable for a paved garden are:-

1. ***Achillea rupestris* (Compositae):-** A mat-forming, tufted plant with white –yellow flower heads.
2. ***A. tomentosa* (Compositae):-** A compact, mat-forming plant with bright-yellow flower heads.
3. ***Alyssum montanum* (Cruciferae):-** A dwarf (7-8 cm), prostrate, tufted plant which bear many-flowered recemes with bright-yellow and fragrant flowers.
4. ***Dianthus deltoids* (Caryopyllaceae):-** Commonly called 'Maiden Pink' is a prostrate dwarf (12-20 cm) herbaceous plant, bearing purples to crimson-red flowers with dark strips and bright spots.
5. ***Geranium oxonianum* (Geraniaceae):-** A slender, dwarf (30 cm tall) plant bearing purple flowers.
6. ***Hymenatherum tenuifolium*:-** A herbaceous perennial or annual of prostrate habit, the fern-like foliage is finely divided and beautiful. Flowering is round the year.

7. *Lantana sellowiana, Gysophila repens, Porulaca* (Perennial type), *Setcresea palida* and *S. purpuria, Verbena erinoides, Vinca rosea, Viola cornuta, Zebrina pendula, Euphorbia splendens prostrate.*
8. ***Zinnia linearis*:** A gardener should be able to add many more plants by practical experience.

Chapter 12

Landscape Designing of Bog Garden

- An actual bog garden is an area where there is stagnant acid water and only plants such as sundew, butter words, etc., grow. But the bog garden which we find in landscape gardening is actually a marsh where the soil is not sour or acid and a shallow stream or trickle runs through it. Therefore, it is also called as *Marsh garden*.
- A pond is not an essential feature of any bog garden, but, if there is any adjacent pond, the overflow water can be used to keep the marsh garden moist.
- The main criterion of a marsh garden is to keep it moist and in a swampy state all throughout. For this reason, the site of the marsh garden should be low-lying where the surface drainage water will collect naturally.

Site selection

- A site having a sub-soil of sticky clay is ideally suitable for marsh garden as only a trickle of water will keep this wet. But, if the soil is light in nature, certain amount of digging will be needed before establishing a marsh garden.
- Construction procedure
- The whole area intended to be developed into a marsh garden is dug up to a depth of 60 cm. and a thin layer of clay, 10-12 cm deep, is placed at the bottom to form the base of the marsh garden.
- Then a 12 cm layer of rubble or large stones are placed over this to ensure that the water in the marsh does not become stagnant.
- The rest of the hollow is filled with a compost of garden loam and leaf-mould in the proportion 1: 1 and the level made up with the surrounding land.
- A marsh garden should never exceed a depth of 60 cm. although its spread will depend upon the available area and personal choice.
- A marsh garden should never suffer from drought and be kept sufficiently moist by a trickle of water at the same time avoiding stagnation.
- All parts of the marsh garden is made accessible by laying paths of rough stones or bricks, winding round the marsh and putting stepping-stones over them.

Selection of Plants

- It must be noted that all marsh plants need water, but in varying degrees. Accordingly, marsh plants can be categorized into three broad groups:
- *Group I-* In the first group belong plants which over 5-10 cm of standing water and are termed as *"Swamp dwellers"*.
- *Group II-* In the second category are placed the plants which live on or a little above the waterline, with their roots standing in water-logged soil.
- *Group III-* This category consists of those plants which grow above the saturated soil, but near enough to send their roots in search of water.
- A marsh garden can accommodate any moisture-loving plant, those which are found commonly growing on the banks of streams or ponds. Some of the plants which can stand in 12-15 cm. of water are also included in a marsh garden. In some pockets of the marsh even plants growing in shallow water can be accommodated.

Plants Suitable for Bog Garden are as follows:

Acorus calamus Calamus var. variegates, Sweet flag, Alocasia, Arundinaria (Bamboo), *Arundo donax,* Fern, Double Marsh Marigold, *Cypripedium spectable,* Day-Lily, Hosta sp., Plantain-Lily; trees such as Acacia farnesia, Barringtonia and Salix can be grown in the background.

Chapter 13

Landscape Designing of Public Places

Historical Places

The places of historic importance such as tomb, forts, historic monuments etc., can be laid out with vast stretches of lawns, large ponds, Canna beds, shade and flowering trees etc., keep in view that the planting should not over-shadow the monument. The trees should be selected depending upon the climate of the region. In dry area, the approach roads to the historic spot can be planted with drought withstanding flowering trees such as *Erythrina indica.*

Religious Places

Temples, Gurdwaras, Mosques, Churches etc., are the places of worship. In such places, the scope is very limited especially in large cities because of space problem. But in villages and towns, the places of worships form good scope for bio-aesthetic planning. It is important to have a garden with the right type of flowering plants.

Buddhist temples: The association of *Plumeria alba* is so common with Buddhist temples that it is commonly called as 'Temple tree' or 'Pagoda tree'.

Hindu temple: Shrubs such as Jasmines, Crossandra, Barleria etc., are quite useful in a Hindu temple. Seasonal flowers such as marigold and small flowering chrysanthemums (yellow flowering type) can also be planted around Hindu temples. Generally, there will be come water tank in the vicinity of a Hindu temple. This should be planted with water Lilies and Lotus and on the banks moisture loving trees may be planted. Some sweet scented flowering trees such as *Michelia champaka* should be planted around the temple to create an atmosphere of serenity and sanctity. There should be enough open space in the front to the devotees to assemble there under the shade of trees. A tree of *Saraca indica* is another sacred tree which should find a place in a Hindu temple.

Church: Here foundation planting with shrubbery is quite useful. Potted plants can also be arranged near the portico. There is scope for displaying hanging baskets especially for Gothic type buildings.

Mosques: In front of Mosques, Mughal type gardens can be laid out with stress on formal water pools fitted with fountains.

Public Buildings

- The government and private offices, courts, auditoria, cinemas, hotel and travelers bungalows fall under the group of public buildings. If there is paucity of space, potted plants are used to decorate the entrance and the corridors.
- But where space is available, a lawn should be laid with a few flower beds and possibly some shrub border may be added.

- The entrance and exit roads may be lined with flowering trees. Majestic looking non spreading large trees will be the right choice for skyscrapers.
- Some flowering climbers may be trained over the portico. If the compound is properly planted with a few shade and flowering trees and a nice lawn, the people will find a resting place.
- A miniature garden with lawn, flower-beds, shrubbery, lily pool, fountain and a few specimen trees can be planted in Travelers' Bungalow garden.
- Hotel buildings should also have gardens of the same style if space is available. Fountains fitted with colored light can also be added. Sufficient parking space should also be provided.

Educational Institutions

- In the periphery of the School Campus, along the rear and wings, a thick beet of large shady trees should be planted with medium sized flowering trees for beauty.
- A flower parade can be created along the periphery. Eucalyptus, silver oak and *Polyanthia longifolia* are suitable for peripheral planting.
- Where are electric wires limit the choice of avenue trees, small flowering trees such as *Cochlospemum gossypium, Callistemon lanceolatus, Bauninia variegata and Tabubia argentia* can be planted *Cassia fistula, C. nodosa, Gulmohar, Tabubia argentea, Erythrina indica, Lagerstroemia flosreginae* and *Bauhinia variegata* are suitable for planting in the front row of the border planting.
- Border planting of shrubbery on large ground or at the back of the school campus serves useful purpose of filling the gaps between the trees and lawns.
- Small paths can be planted with shrubs. Shrub borders around playgrounds are very effective. The foundation planting can also be undertaken with evergreen dwarf and trailing shrubs.
- A playground can be planted with lawn. A bougainvillea creeps, trained over the wall of the building or *Bignonia venusta* supported against a wall looks beautiful.
- Creepers such as *Ficus repen, Ficus puemilla, Campsis radicans (Syn. Tecoma radicans)* can be trained over some stone or brick wall or pillar.

- A bird bath provided amidst the trees will be educative to the students.

Factories

- In factories such as cement, steel, fertilizers etc., which emit a lot of dust, smoke and harmful chemicals, and comparatively hardy plants, are to be selected.
- For neat factories such as a plywood factory and a fruit-processing plant, which emit less dust and other polluting materials, a wider range of plant which emits less dust and other polluting materials can be used.
- The primary aim in a factory garden is to plant trees to arrest the drifting dust and smoke and to cut down noise.
- Tall and hardy trees such is *Casuarina equisetifolia, Eucalyptus, Polyalthia longifolia* and silver oak should be planted all round or in the direction of the winds to stop the spread of dust and smoke.
- In a large factory a buffer zone may be created by forestation between the factory and its residential colony.
- Afforestation can be done with hardy ornamentals such as *Acacia auriculiformis, Casuarina equisetifolia, Dulbergia sisso* and some other shade trees.
- A factory can also be beautiful with rockeries, statues, water pools or lakes, fountains etc. Bougainvillea should be used freely to beautify a factory area. This may be planted in rows along all the roads of the factory spaced at 3-5m and trained as half standards.
- Canna a hardy plant is suitable for planting in beds or groups in a factory area.
- Shrubberies consisting of hardy shrubs also beautify factories.
- Lawns add beauty and cut down dust.

Landscaping of Highways

- The landscaping of the National and State highways with trees is an important aspect of beautifying our countryside. Planting of trees along the roadside is known as avenue plantation.
- Planting of trees on highways is necessary not only for the purpose of beautification but also for utility and necessity. The main purpose of roadside trees is to provide shade during the summer. For this purpose, evergreen trees with spreading crowns should be selected.
- For wider roads, double rows can be planted, with the outer rows having shade trees and the inner rows with flowering trees.
- The planting of roadside trees started during the time of Emperor Asoka (268-231 B.C.). The Mughals also planted roadside trees.
- The roadside trees on the national highways not only provide shade but preferably also have some economic value. With this in view many of the highways have been planted with trees such as tamarind, mango, *Eugenia operculata* (Syn *Syzygium operculatum*), etc.
- *Eugenia operculata* has one nuisance value as when its fruits ripe they drop on the road and make it dirty. The stain of this fruit is also difficult to remove from the clothes. Neem (*Azadirachta indica; Madhuca indica; Dalbergia sisso* (Sisham), and *Shorea robusta* (Sal) are roadside trees of economic value.
- The highway trees should never be planted in mixed avenues, but only one species should be planted for a long distance of the road. If a single species is planted in a pure avenue for miles together this looks more beautiful and gives a wavy appearance to the skyline
- The trees should be planted 12 m apart in the row and at least 5-6 m away from the edge of the roads, so that they get enough space for spreading and do not interfere with the traffic. If a road is as wide as 30 m or more, double rows of trees should be planted, rows being spaced 10-12 m apart. The inner row may be of a flowering tree.
- Trees with shallow root system such as *Millingtonia hortensis* and brittle wood as in the case of *Eugenia jambolana, Albizzia lebbek, Cassia siamea,* and Eucalyptus should never be planted on highways, as during storms they get uprooted or branches are broken and casualties may result on the unaware road users, Neem and Tamarind can grow very well in dry localities.
- *Samanea saman* and *Dalbergia sisso* grow better in places having a rainfall of 100 cm or above. Banyans can be planted singly and a little away from the road at some distances, for its cool shade. *Albizzia procera* is a good roadside tree.
- *Polyalthia longifolia,* though not a tree of great economic value, is a very good shade tree and thus suitable for roadside planting. Thorny trees should not be planted along roadside as the falling thorns may damage the tyres of vehicles. Shrubbery borders with suitable shrubs can be maintained along the highways. This will improve the scenery.

Landscaping of Railway Stations and Railway Lines

- A well-kept and well-planted railway platform gives a visitor or a passer by the first impression about the town. It should be a social obligation on the part of our railway authorities to keep the platforms planted with beautiful trees.
- Trees offer the passengers a resting place under the hot sun, while waiting for the train. Concrete seats or benches could be constructed around the trunk or under the canopy for the benefit of the commuters.

Railway Stations

- Besides flowering and foliage trees, the railway authorities can also improve and beautify the platforms with tubs and troughs planted with palms and other attractive plants such as bougainvilleas.
- Even hanging baskets can be displayed near the booking office or on the pillars of resting sheds and in similar other places. The railway stations are landscaped with avenues of such flowering beauties as *Cassia fistula* (amaltas), *Cassia nodosa, Peltophorum ferrugineum, Delonix regia* (gulmohar), etc.

Railway Line

- It will be a little hard task to landscape the vast stretches of railway lines compared to landscaping the platforms, especially in the drier tracts of the country. Fortunately, in India we have got beautiful flowering trees such as *Butea monosperma* (Flame of forest/Palas), *Cassia fistula, Erythropsis colorata*, etc., which can withstand considerable drought conditions.
- The landscaping of railway lines poses some problems, the basic being watering. This can be partly overcome by planting drought-resistant trees during the rains. Plants should be deep-rooted and non-spreading. The trees are to be planted at a specified distance from the tracks as per railway regulations. These plants, besides breaking the monotony and drabness of the landscape, will offer shade to the gang workers.
- The innumerable railway level-crossing are the places, which can be landscaped with much ease than the railway line, as the gate-man will be there to look after the plants.
- The gate-men who live there in the houses provided for them can be asked to plant a row of a few flowering trees on either side of the lines on both sides of the gate or in a small group-near the gate.

- The beauty of such trees will not only be enjoyed by the railway passengers but also by the people who pass by the road or wait for the train to pass.

  Trees suitable for planting on railway platforms and lines:

  1. **Ornamental purpose:** *Albizzia procera, Bauhinia variegata, Cassia fistula, C. javanica, C. nodosa, Delonix regia, Gliricidia maculate, Jacaranda mimosifolia, Lagerstroemia spp, Michelia champaka, Peltophorum ferrugineum, Polyalthia longifolia* and *Saraca indica.*
  2. **Economic purpose:** *Anacardium occidentale, Averrhoa carambola, Dalbergia sisso, Mangifera indica, Melia azedarach, Shorea robusta, Swietenia mahagoni, Tamarindus indica, Tectona grandis, Terminalia arjuna* and *T. catappa.*

Landscaping of Bus Terminus and Airport

- Now-a-days, with the increase in road traffic many State Governments are constructing large bus terminals and bus stations. The bus terminals should be beautified on the lines of railways stations.
- Airports should also be beautified by planting ornamental trees, lawns, flower beds and displaying plants in tubs and troughs. Fortunately, many of the airports in India are quite nicely landscaped, but there is some scope to improve them.

Landscaping along Banks of Rivers and Canals

- The banks of the rivers and canals passing through towns and villages are well-suited for landscaping. The plants themselves will be objects of beauty and their reflection on the water surface is an additional attraction.
- The moisture-loving trees will do better along the water-front compared to others. In India, the rivers Jamuna, Ganga, Kaveri etc. are regarded as sacred and on the banks of these rivers are situated many old Hindu temples and ghats.
- The banks along these places should be planted with flowering trees, especially the sacred trees such as Kadamba (*Anthocephalus cadamba*) and Asoka (*Saraca indica*) which are associated with Lord Krishna and Sita, respectively.
- The bank of the river Hoogly on the Kolkata side has been beautifully landscaped by the Kolkata Port Commissioners with shrubs and trees and has become a place for recreation for the citizens of the metropolis who can breathe fresh air away from the congested city. Besides, beautification, planting of trees along banks of rivers and canals help soil conservation also.
- The dam sites which generally look drab should be converted into a place of beauty by landscaping with such beautiful trees as *Cassia nodosa, C. javanica,* Gulmohar, etc. A garden or park can also be planned here, if space permits. The Brindavan Gardens in Mysore constructed below-a-dam site has become a place of great tourist attraction.
- River banks near places of historic importance, such as Taj on the bank of Jamuna, should also be landscaped with beautiful trees.
- ***Flowering trees suitable for planting along the banks of rivers and canals:***
- *Amherstia nobilis, Anthocephalus cadamba, Barringtonia acutangula, Bauhinia variegata, Browneas, Cassia fistula, C. javanica, C. marginata, C. nodosa, Delonix regia, Largerstroemia flos-reginae, L. thorelli, Melia azedarach, Peltophorum ferrugineum, Samanea saman,* and *Saraca indica.*

Landscaping of City, Town and Countryside

In India most of the old cities and towns are very much unplanned without any consideration given to landscaping with trees and plants and as a result they look nothing but like brick and concrete jungles.

- The vast majorities of our population in the congested cities are cramped inside the four walls of their dwellings and are constantly injected with the fumes of petrol and other polluting agents.
- Our towns and cities can be made more livable, healthy and beautiful and resorting to bio-aesthetic planning. This can be achieved by planting roadside trees and establishing parks planted with green plants.
- In the old congested cities this can be achieved by planning garden suburbs. Garden suburbs are nothing but satellite townships developed among parks and avenues.
- The cities can further be beautified by accommodating the slum dwellers in multi-storey buildings and utilizing the surplus land thus obtained for developing parks.
- The planning of new towns poses no problems as this can be planned in advance even before construction starts.
- In a tropical country like India, the planting of roadside avenues are not only important for beautification, but also from utility point of view.

Selection of trees for cities and towns with main criteria should be beauty and shade:

- The broad roads in the cities should be planted with double rows of trees; the outer rows should consist of shade trees, so as to provide shade to the footpath users and the inner row consisting of flowering trees adding beauty.
- M.S. Randhawa suggested long back that wide roads in the cities and towns should have two lanes on each side, one for fate-moving vehicles and the other for slow-moving vehicles flanked by footpaths. Each lane should be separated by a patch of land planted with grass and shrubs, while the outer rows are planted with double rows of flowering and shade trees.
- Planting the roads in a town or a city is a difficult proposition and this should be done in coordination with the electricity, telephone, and sewerage departments.
- The tree should not interfere with the telephone or electric lines of the underground sewers.

- Tall trees (Eucalyptus, Araucaria, *Millingtonia hortensis*) and trees with spreading habit (Banyan, *Cedrus deodara, Ficus benjamina,* etc.) are not suitable for city or town roads.
- The tropical flowering trees are generally deciduous in habit and flower in leafless condition, which may be at a time when shade is needed to the maximum. The solution is, wherever feasible, to have an outer row of small shade trees such as *Ficus infectoria, Putranjiva roxburghii,* Silver oak, *Mimusops elengi,* etc., and inner row of flowering trees. The trees in either the rows, or when there is only a single row, should be planted in pure avenues, i.e., with one kind of tree only.

In many cities a piece of land is left out at the intersection of roads as a safeguard against accidents. These plots should be planted with grass and shrubs. One more novel way of beautifying the city roads is to train creepers on iron structures at regular intervals on foot paths.

Landscaping of City Parks

- In city there may be parks of several sizes from very large to medium size and also squares or small gardens are generally found at street intersections. The small gardens or squares are planted with a view to relieving the eyes of the people passing by them or for a short resting period for shoes who care to use them.
- Therefore, these may be planted with a patch of grass, few flower beds, one or two shade or flowering trees or a group of shrubs and trees. The medium to large parks are meant for a place of recreation and these are considered as lungs of the cities. These should be a place of beauty as well as utility.
- Good flowering and shade trees should be planted in groups or singly in some corners or other suitable places for creating beauty as well as a place for resting. Garden benches should be constructed at regular intervals especially under the shade of the trees. Few interesting and rare shrubs should also be included. Besides these, some garden adornments such as statues and fountains can also be planned in appropriate parts of such parks.
- City parks may be called as pleasure grounds which have large reserve areas for playing games, and often this is the main feature of these parks. A restricted swimming pool is also often a feature of a pleasure ground. If it is meant for the children, features such as swings, see-saw, sliding chute, merry-go-round, etc., should form part of the park.
- Due to hard usage it is almost impossible to maintain grass area within this park. Some trees and shrubs are planted aesthetically to keep the place pleasing to the eyes.

Trees Recommended for Town Roads

- **Ornamental Shade Trees:** *Anthocephalus cadamba, Ficus infectoria, Melia azadirachta , Polyalthia longifolia, Putranjiva roxburghii, Swietenia mahagoni, Tamarindus indica.*
- **Flowering Trees:** *Bauhinia purpurea, B. variegata, Cassia fistula, C. javanica, C. marginata, C. nodosa, Gliricidia maculate, Jacaranda mimosaefolia, Lagerstroemia flos-reginae, Peltophorum ferrugineum, Poinciana (Delonix) regia, Saraca indica,* and *Spathodea campanulata.*

Landscaping of Country Side

- The planting of trees in the countryside is done on a completely different criterion. Here more emphasis should be given on economic consideration and utility. The trees planted should provide the villager either with fuel, timber, fruit, or fodder.
- Babul is a common tree in the village, which is a cheap source of fuel and is used for the manufacture of wheel of bullock-carts. The bark of this tree is used for tanning leather.
- Desi (seedling) Mango and Jackfruit trees also yield timber for making doors and windows and other household articles.
- *Dalbergia sisso* is also a good source of timber and fuel.
- Sal (*Shorea robusta*) trees are good for both fuel and timber.
- Some minor fruit trees should be planted along the country road in open spaces. This will not only provide shade but will yield valuable fruits for the villagers.
- Jackfruit is an important fruit tree for arid regions.
- The Toddy or Palmyra palm (*Borassus flabellifer*) is an important tree, which yields a minor edible fruit but the leaves are used extensively for thatching roofs of dwellings.
- Both Date and Palmyra palms look beautiful when planted in rows along country roads or on the borders of cultivated fields or when planted in informal groups.
- Mahua and large-fruited Jamuns (*Syzygium cumini*) can also be planted.
- The foliage plants yield valuable fodder for the cattle and goats

- Babul, Kachnar (*Bauhinia*), neem, and Ber (*Zizyphus species*) also produces edible fruits.
- However, ornamental trees should also be planted near the Panchayat, Schools, Churches, Mosques, Temples and Gurudwaras and along the field boundaries and near the village well or pond.

Landscaping of a Country Home

- A villager in India may not need a sophisticated garden. However, a countryman will need as much privacy as a city dweller. A village home has to be planned with more utility items.
- A villager would like to grow more vegetables and fruit for his family consumption and consequently more area should be earmarked for this purpose. But few shade trees and some area reserved for children's play-ground is definitely needed.
- Some utility *flowering trees* such as *Michelia champaka, Plumeria acutifolia* and shrubs such as *Hibiscus rosa-sinensis, Tabernaemontana coronaria, Barleria, Crossandra,* and jasmines yielding flowers for worshipping and hair decoration should be included for planting.
- The common *flowering annuals* such as marigold, zinnia, balsam, and sunflower can be grown for beauty and cut flowers for various purposes.

Landscaping of Cemeteries and Burning Ghats

The landscaping of cemeteries and burning ghats may sound paradoxical under Indian conditions. This is a country where a large number of people do not even get the minimum amount of food to keep them alive during their life-time, so the idea of having a garden around their tombs after death may sound ridiculous at the first thought.

- But there is a need for this also, not only from the point of peace, tranquility, and beauty, but also to fight pollution and provide shade. Whether the body is buried or burnt in an ugly surrounding or a beautiful landscape matters nothing to the departed person (unless, of course, we believe in life after death), but it certainly has some sobering effect on the sorrow-laden hearts of the near and dear ones who attend the funeral. While bidding final farewell one would definitely like the last resting place of one's dearest in a quiet, peaceful surrounding. For this reason also, a cemetery needs to be landscaped with trees, shrubs, and grass.

- The cemetery is sub-divided into several sections and each section is made accessible by providing driveways at a distance of 50-100 m depending upon the size of the place. In small places of 3-4 hectares no driveway is needed and the walks in such cemetery may be left in grass only. Most of the area in the cemetery should be left under grass with few trees grouped here and there for immediate effect. More trees, especially the evergreen types, are planted along the boundaries. Some flowering trees etc, of course, selected to bring a bright and cheerful effect. The buildings should be modest in appearance and be planted with shrubbery and vines.
- Under Indian conditions also it is possible to keep the cemeteries in a park-like fashion without much elaborate arrangements. The basis will be the same as stated above with emphasis on modesty and sobriety. In India many cemeteries have been planted with trees mainly for the purpose of shade and to create some formal effect. Very often there will be some shade-trees such as seedling mango, banyan, neem, etc. and the roads are lined with *Polyalthia longifolia* var, pendula, *Plumeria acutifolia* may also be planted in burial grounds, which flowers freely under most Indian conditions.
- The burning ghats in India present a picture of horror and sorrow, which can definitely be improved upon by proper landscaping. It is not possible to have a formal garden with grass, flower beds, etc. in a conventional burning ghat, for obvious reason, but the picture can be changed by planting in groups, some shade and flowering trees. In most cases these ghats are located near water tanks, ponds or rivers. The ponds can be planted with water-lilies and lotus while the banks can be beautified by planting shade and flowering trees. In modern electric crematoriums it is possible to layout a formal or an informal garden around it with important features such as trees shrubbery, flower beds, lawns, etc.
- The "*Towers of Silence*" of the Parsi community affords good scope for gardening. Since these towers have long stretches of ground all around them, a nice garden with lawn, shrubbery, and flowering trees along the periphery, can be laid out. For instance, the Tower of Silence near Bangalore has a long avenue of jacarandas which, when in bloom, present wonderful scenery.

Chapter **14**

Landscape Designing of Home Garden

- Landscaping as it is done for larger estates or public parks can also be implemented in a tasteful and artistic way for a small home ground, though on a smaller scale.
- The simplest definition of "small", as suggested by some authors quite appropriately, is an area which can be effectively managed and maintained physically as well as financially by the owner and his family with occasional hired labour for such hard work as digging, mowing, and shearing of hedges.
- Here, ways will be suggested for landscaping only small residential houses. For larger estates, a combination of landscaping effects suggested for parks and home landscaping may be followed.

Basic Guidelines for a Home Landscape

- The personal preference plays a considerable role in developing a home garden.
- The home including its surroundings should be an outward expression of the inner personality and individuality of the owner.
- It is advisable to think a lot before even a single digging work starts.

Basic Principles: Before actually drawing the master-plan the following points should be kept in mind in home landscaping:-

1. **Background:** The background in a garden, whether a wall, tall trees or a hedge should be neutral in nature; that is to say this should not become a distracting feature over the main features of the garden.
2. **Contrast:** The design should be such that it should break monotony. To achieve this, variation in form, texture, or colour has to be brought in.
3. **Balance or Proportion:** A balance has to be maintained between different components (masses, forms, colour, etc) of a garden. Overcrowding of plants or other garden features should be avoided.
4. **Open Centre:** The central area of the garden should be left out of any items of major interest. The best way to achieve this is to have a lawn, which also gives an effect of largeness to the property. A specimen shrub in the centre of the lawn is unsuitable as this counters the principle of spaciousness, but a tree branching at higher levels from the ground could be planted.
5. **Repetition:** The repetition or duplicating some features of a garden helps achieve rhythm, balance and unity. In a formal garden, generally the same feature is repeated. But for an informal design this need not be so. Here one may repeat the colour tone without disturbing the texture, form, or quantity. If there is a circular path this can be repeated by having two or three consecutive circular shaped beds of annual flowers, hedges, and shrubbery border. Thus, though the shape is repeated, the variation in texture, colour and form ensure that the design does not look monotonous.
6. **Rhythm:** A landscape designer should have an artistic sense to understand how to bring in rhythm in the design. Arranging the different elements haphazardly, without harmony, does not enhance beauty. Harmonious lines, often artistically curved, bring in rhythm to the landscape. A group of shrubs in front of a rockery breaks the rhythm. Repetition of certain elements, such as form, enhances the rhythm.
7. **Variety:** To break the monotony in a garden, variety is essential. This is achieved by contrast of colour, form, and texture. Besides these, design should be simple, easy to maintain and provide comfort for inmates.

Making a Plan- If the garden area is sufficiently large, this can be divided into three areas.

1. **Approach or Public Area**
   - This is the area from the street side extending to the entrance of the house. The approach area should not be overcrowded with large trees. It is better to have doorway or "foundation" plantings with low growing shrubs and evergreens.
   - Big trees, if space permits, can go in the backyard but should not be overcrowded in the front. But a few low-growing trees can be accommodated at the appropriate places as next to entrance, if space is available or somewhere in the front lawn.
   - An open spacious lawn with some annuals (zinnias, salvias and petunias) or herbaceous perennials (chrysanthemum, canna and Impatiens in shade) can be planned in addition to the foundation plantings.
2. **Work or Service Area**
   - Wherever feasible this and the living area should be situated at the back of the house as these need privacy.
   - This area includes the kitchen garden, compost bin, nursery, tool shed and garage.
3. **Private Garden Area or Living Area**
   - In the western countries, this is generally termed as the outdoor living area, where people sit out in the winter to enjoy the sun or rest in the summer under an arbour or shade of tree.
   - This area should be easily approachable and visible from the living (drawing-room) or dining-room, screened from unsightly objects and for privacy. In the western countries people prefer a terrace and this is the place where it should come.
   - There should be some shaded sitting spot such as a tree or arbour with garden benches. A wide stretch of lawn with shrub border or few annuals beds or a rose garden can also be included in this section. A tennis court or a play area has to be included here, if there is enough room.
   - A doorway near the house needs special attention as this is the place which receives maximum attention from a visitor. Depending upon the approach a doorway can be planted informally, formally, or in a semi-informal pattern.

Few Examples of Home Gardens

- Many people advise not to include any pool or formal rock garden or the kind in a home garden. But there is no harm if a formal or informal lily pool can fit in with the overall design, with or without a fountain or a rock garden. A statue or sun dial can also be well fitted in some spacious compounds.
- It can be arranged with a garden-type design consisting of an ornamental shade or flowering tree perennial and annual flowers, climbing roses, some bulbs such as Zephyranthes, Amaryllis, and daffodils (for temperate regions).
- A bed of roses can also be a spot of beauty provided it receives the morning sun. Symmetrical plants with pyramidal form such as Thuja, Juniperus chinensis, and Cupressus macrocarpa are preferred by many near the doorway for a formal treatment.

Trees Suitable for Small Gardens

Bauhinias in different species are quite suitable. Bottle brush is suitable for many situations. *Tecoma argentea, Cassia fistula*, and *Cassia spectabilis*, Cherries, (*Prunus sargentii*), Weeping willow (*Salix babylonica*) and *S. purpurea var. pendula* are also very ornamental. Some shrubs *Ixora singaporensis, Mussaenda philippica, Azalea* etc.

Designing of Greenhouse, Conservatory and Lathe House

- **A conservatory** is a glass and metal structure traditionally found in the garden of a large house or public park. Modern conservatories are smaller, can be made of PVC and are often added to houses for home improvement purposes. Whereas, **a greenhouse** is a structure with a glass or plastic roof and frequently glass or plastic walls; it heats up because incoming solar radiation from the sun warms plants, soil, and other things inside the building faster than heat can escape the structure.

- Air warmed by the heat from hot interior surfaces is retained in the building by the roof and wall. These structures range in size from small sheds to very large buildings. A garden conservatory is usually a small conservatory usually attached to a private house.
- In this context, a conservatory is distinguished from a greenhouse in having a role as a living space as well as being used to grow plants. If the space is not used to grow plants is it best described as a **solarium.**
- There is actually very little difference between a greenhouse and a conservatory structurally but these differ as to the use to which they are put.
- Originally and etymologically the term greenhouse meant a house in which plants are kept there to grow. But sometimes a green house may be used for a different purpose also. For example, in severe winter many plants are kept inside for winter protection only and are not likely to grow.
- The Americans use the term greenhouse to mean any type of glass building inside which plants are grown, except the structures such as cold frames and hotbeds.
- In England, the term is defined as a glass structure in which plants not requiring a high temperature are cultivated or displayed. In a greenhouse, plants are exclusively grown in pots and tubs.
- According to the British terminology the term has wide application as it may include houses in which plants are grown from seed to its maturity, or may mean a house exclusively used for displaying plants which were grown up to the flowering stage in another house.
- A conservatory differs from a greenhouse in which permanent plants are planted in central beds and the side beds are used for continuous floral display. A conservatory is situated very near the dwelling house and should actually be included in the same architectural plan. This may be connected with the house by a corridor or a pergola.
- In India conservatory is also known as **Fernery** as in olden days ferns occupied predominant position in such houses.
- It will be seen from the above discussion that there is no one terminology which will include all the glass plants houses. Some people may use the term glasshouse to include these but there are other glasshouses also in which plants are not grown.
- Hence, the best terminology is greenhouse meaning a structure use d for growing living or green plants.
- Under Indian climatic conditions, the term greenhouse includes a glasshouse for plants and any other structure including Lath house for growing shade loving or greenhouse plants. In the tropical plains of India hardly any glasshouse is used for growing plants except for researches on plant viruses. Even if glass structures are made say for growing cactus and for breeding purposes it is generally only the roof which is made of

glass and the sides are covered with fine wire-mesh. Otherwise, a full glasshouse has to be air-conditioned to keep it cool during the summer months.

- In most parts of India a greenhouse and conservatory or fernery, as it is called popularly, is meant for providing shade and cool temperature for the plants growing within it. A greenhouse provides a cool and pleasant retreat for the dwellers and friends in the hot summer. In high hills, a glass greenhouse is necessary to grow the more tropical plants to protect them from severe cold weather.
- The glass used for a greenhouse works as a selective transmission medium for different spectral frequencies, and its effect is to trap energy within t he greenhouse, which heats both the plants and the ground inside it. This warms the air near the ground, and this air is prevented from rising and flowing away. This can be demonstrated by opening a small window near the roof of a greenhouse: the temperature drops considerably. This principle is the basis of the auto vent automatic cooling system.
- Greenhouses thus work by trapping electromagnetic radiation and preventing convection. A miniature greenhouse is known as a **cold frame.**
- **Lath house** is a valuable asset in raising seedlings, rooted cuttings, and young rhododendron plants prior to setting them out in the garden. Properly used, it will modify the environment in which they are growing by offering protection from hot, drying summer winds, reducing the intensity of the sunlight, lowering temperatures, and by maintaining a higher humidity. During December, January and February, the sides may be covered by plastic sheeting, which will furnish additional protection from the winter cold.
- Under Indian climatic conditions, the term green house includes a greenhouse for plants and any other structure including lath house for growing shade loving plants.

Index

E

F

G

H

I

J

L